THE GENERALIZED JACKKNIFE STATISTIC

STATISTICS

Textbooks and Monographs

EDITED BY

D. B. OWEN

Department of Statistics
Southern Methodist University
Dallas, Texas

VOLUME 1: The Generalized Jackknife Statistic, *H. L. Gray and W. R. Schucany*

VOLUME 2: Multivariate Analysis, *Anant M. Kshirsagar*

OTHER VOLUMES IN PREPARATION

The Generalized Jackknife Statistic

H. L. GRAY
Department of Mathematics
Texas Tech University
Lubbock, Texas

and

W. R. SCHUCANY
Department of Statistics
Southern Methodist University
Dallas, Texas

1972

MARCEL DEKKER, INC. New York

MARCEL DEKKER, INC.
95 Madison Avenue, New York, New York 10016

LIBRARY OF CONGRESS CATALOG CARD NUMBER: 75-179385
ISBN: 0-8247-1245-5

PRINTED IN THE UNITED STATES OF AMERICA

To
Becky and Carol

PREFACE

In the last several years there has been an increasing interest in the development of robust statistical methods that do not depend on the assumption of normality, but that retain many of the desirable features afforded by this assumption. In connection with this problem, the advent of the digital computer suggested a number of possible procedures which previously had to be considered impractical but which now seem quite reasonable. Among these procedures is a method commonly referred to as the jackknife method, which is the subject of this book.

The jackknife is a general method for reducing the bias in an estimator and for obtaining a measure of the variance of the resulting estimator by sample reuse. Thus the result of the procedure is usually a nearly unbiased estimator and an associated approximate confidence interval.

It is the purpose of this book to explore to some extent the theory surrounding the jackknife method and to exemplify its applications. Since this theory is in its early stages of development, it may be that more questions are asked than answered. On the other hand, in spite of the many interesting questions that arise in connection with the technique, a large body of theoretical results and practical examples have already been established. Moreover, many of these results appear to be of some importance, and since they can only be found scattered among the various journals (at least part of them can be found in the literature) this collection should be helpful.

An attempt has been made by the authors to include most of the current results, or at least a reference to them, and to give a rather uniform theory of the subject. In order to do this it was often necessary to give some additional developments and as a consequence the book is sprinkled with a number of new results, which we believe add to its cohesiveness. In spite of our efforts to cover the field thoroughly, we have undoubtedly missed some things we should have mentioned. For this we apologize and, we might add, look forward to hearing about it (as we surely will).

Although the book is in the nature of a research monograph, it is written in a style which should allow it to be useful to the practitioner who is uninterested in the details of the theory. Many examples are included and the meaning of a theorem is usually discussed. It is our sincere hope that both statisticians and users of statistical methods will find this work interesting and helpful.

The authors would like to express their appreciation to the Office of Naval Research and Dr. Bruce McDonald, who sponsored several invaluable research conferences on this topic under ONR Contract N00014-68-A-0515. Professor Donald B. Owen is gratefully acknowledged for his continuing encouragement and for making this book possible. Also we are deeply indebted to Dr. T. A. Watkins and Dr. J. E. Adams for their generous contributions to this book and for their careful reading of the manuscript. Dr. James Minor is also acknowledged for his assistance with the Monte Carlo studies. Finally we would like to express our gratitude to Miss Alonda Smith, Mrs. Myra Archer, and Mrs. Kathy Williams for their patient typing of several versions of the manuscript.

Lubbock, Texas
Dallas, Texas
January 1972

H. L. Gray
W. R. Schucany

CONTENTS

CHAPTER I
REDUCTION OF BIAS BY JACKKNIFING

1. Introduction

In this chapter we will define and investigate the bias reduction properties of an estimator which we shall refer to as the generalized jackknife statistic, or simply the jackknife. A version of this estimator was first introduced by Quenouille [34] for the purpose of reducing bias, and this version was later utilized by Tukey [46], to develop a general method for obtaining approximate confidence intervals. At that time Tukey referred to his method as the "jackknife." Since then a number of papers have been written on the Tukey method and Quenouille's estimator. They have proven their worth as useful tools for the applied statistician, and the method as well as the estimator are now generally referred to as the jackknife. In the definitions below we shall essentially follow the line of development given in [44].

2. The Generalized Jackknife

DEFINITION 2.1. Let $\hat{\theta}_1$ and $\hat{\theta}_2$ be estimators for θ. Then for any real number $R \neq 1$ we define the generalized jackknife $G(\hat{\theta}_1, \hat{\theta}_2)$ by

$$G(\hat{\theta}_1, \hat{\theta}_2) = \frac{\hat{\theta}_1 - R\hat{\theta}_2}{1 - R} \quad . \tag{2.1}$$

Clearly if the value of R depends on n and $\lim_{n\to\infty} R$ exists and is different from 1, then, when $\hat{\theta}_1$ and $\hat{\theta}_2$ are consistent for θ, $G(\hat{\theta}_1, \hat{\theta}_2)$ is also consistent. A trivial but important property of $G(\hat{\theta}_1, \hat{\theta}_2)$ is given by the following theorem.

THEOREM 2.1. If

$$E[\hat{\theta}_k] = \theta + b_k(n,\theta), \quad k = 1,2 \quad ,$$

$$b_2(n,\theta) \neq 0 \quad ,$$

and

$$R = \frac{b_1(n,\theta)}{b_2(n,\theta)} \neq 1 \quad , \tag{2.2}$$

then

$$E[G(\hat{\theta}_1, \hat{\theta}_2)] = \theta \quad .$$

Proof:

$$E[G(\hat{\theta}_1, \hat{\theta}_2)] = \frac{E[\hat{\theta}_1] - RE[\hat{\theta}_2]}{1 - R}$$

$$= \frac{\theta(1-R) + b_1(n,\theta) - Rb_2(n,\theta)}{1 - R}$$

$$= \theta \quad .$$

The significance of Theorem 2.1 is of course that, if R is known and given by (2.2), $G(\hat{\theta}_1,\hat{\theta}_2)$ is an unbiased estimator for θ. An extension of the above theorem which we shall find useful is as follows.

THEOREM 2.2. If

$$E[\hat{\theta}_k] = \theta + \sum_{i=1}^{\infty} b_{ki}(n,\theta), \quad k = 1,2$$

and

$$R = \frac{b_{11}(n,\theta)}{b_{21}(n,\theta)},$$

then

$$E[G(\hat{\theta}_1,\hat{\theta}_2)] = \theta + \frac{\sum_{i=2}^{\infty} b_{1i}(n,\theta) - R \sum_{i=2}^{\infty} b_{2i}(n,\theta)}{1 - R}. \tag{2.3}$$

From (2.3) we shall see that in many cases, although $G(\hat{\theta}_1,\hat{\theta}_2)$ is not unbiased, it does contain less bias than either $\hat{\theta}_1$ or $\hat{\theta}_2$.

In considering $G(\hat{\theta}_1,\hat{\theta}_2)$ as an estimator for θ when R is known, the question which immediately arises is the manner in which $\hat{\theta}_1$ and $\hat{\theta}_2$ should be selected. From (2.1) one sees that

$$\operatorname{var} G(\hat{\theta}_1,\hat{\theta}_2) = \frac{1}{(1 - R)^2} [\operatorname{var} \hat{\theta}_1 + R^2 \operatorname{var} \hat{\theta}_2 - 2R \operatorname{cov}(\hat{\theta}_1,\hat{\theta}_2)] , \quad (2.4)$$

and hence within the class of estimators for which R is positive and fixed we would desire that $\hat{\theta}_1$ and $\hat{\theta}_2$ have a high positive correlation. On the other hand it would appear that in the set of all $G(\hat{\theta}_1,\hat{\theta}_2)$ one would prefer to have $R < 0$ and $\hat{\theta}_1$ and $\hat{\theta}_2$ negatively correlated. Unfortunately a general method for accomplishing the latter is yet to be established. For the aforementioned case, however, a procedure has been established by Quenouille and we shall describe it shortly. Before proceeding we should, however, lay to rest some questions regarding $G(\hat{\theta}_1,\hat{\theta}_2)$ which may have already arisen in the mind of the reader. First, will R ever be known; and second, if $\hat{\theta}_1$ is a function of only the minimal set of sufficient statistics, will $G(\hat{\theta}_1,\hat{\theta}_2)$ necessarily introduce superfluous variability? The answers to these questions are yes and no, respectively. To see that for a large class of estimators R may be known, consider the following. Let $\hat{\theta}_1$ be a specified estimator on the random sample $X_1,X_2,\ldots,X_n$ with the property that

$$E[\hat{\theta}_1] = \theta + b(\theta)f_1(n) \quad .$$

Now let $\hat{\theta}_2$ be the same estimator defined on the subsample of $X_1, X_2, \ldots, X_n$ obtained by deleting at random some X_i. Then

$$E[\hat{\theta}_2] = \theta + b(\theta) f_1(n - 1)$$

and

$$R = \frac{f_1(n)}{f_1(n - 1)} ,$$

so that when $f_1(n)$ is known R is known. Since in general R will be a function of n we will adopt the notation R(n) for R. To answer the question of induced variability we refer ahead to Example 3.1 of Section 3 where

$$\hat{\theta}_1 = \frac{1}{n} \sum_{i=1}^{n} (X_i - \overline{X})^2 .$$

Selecting $\hat{\theta}_2$ as described above yields R(n) = (n - 1)/n and $G(\hat{\theta}_1, \hat{\theta}_2)$ turns out to be the unique minimum variance unbiased estimator σ^2.

We shall now discuss a general procedure for selecting $\hat{\theta}_2$ when $\hat{\theta}_1$ is given.

3. The Method of Quenouille

Let $\hat{\theta}$ be an estimator defined on the random sample $X_1, X_2, \ldots, X_n$. Now partition this sample into N subsets of size M so that NM = n and form a new random sample by arbitrarily deleting a subset of size M from the original sample. Then we define the estimator $\hat{\theta}^i$ to be the estimator $\hat{\theta}$ defined on the subsample which arises when the ith subset of size M has been deleted. Now let

$$J_i(\hat{\theta}) = N\hat{\theta} - (N - 1)\hat{\theta}^i, \quad i = 1,\ldots,N \tag{3.1}$$

and

$$J(\hat{\theta}) = \frac{1}{N}\sum_{i=1}^{N} J_i(\hat{\theta})$$

$$= N\hat{\theta} - (N - 1)\overline{\hat{\theta}^i} \quad . \tag{3.2}$$

The estimator $J(\hat{\theta})$ is called the jackknife and the estimators $J_i(\hat{\theta})$ are called pseudovalues of the jackknife. Both estimators were introduced by Quenouille in [34]. It is clear that they are special cases of $G(\hat{\theta}_1,\hat{\theta}_2)$ and when we are referring to the particular form of $G(\hat{\theta}_1,\hat{\theta}_2)$ given by equation (3.2), i.e.,

$$R(N) = \frac{N - 1}{N}$$

$$\hat{\theta}_1 = \hat{\theta}$$

and

$$\hat{\theta}_2 = \frac{1}{N}\sum_{i=1}^{N}\hat{\theta}^i = \overline{\hat{\theta}^i} \quad ,$$

we will use the notation $J(\hat{\theta})$. However, we shall often refer to any form of $G(\hat{\theta}_1, \hat{\theta}_2)$ as the jackknife except in particular instances where confusion may arise. In those cases the terminology generalized jackknife will be employed for clarity.

The motivation which initially led to $J(\hat{\theta})$ is as follows. Suppose (as is the case for a large number of estimators)

$$E[\hat{\theta}] = \theta + \sum_{i=1}^{\infty} \frac{a_i}{n^i} = \theta + \sum_{i=1}^{\infty} \frac{a_i}{N^i M^i},$$

where the a_i may be functions θ but not of n. Then

$$E[\overline{\theta}^i] = \theta + \sum_{i=1}^{\infty} \frac{a_i}{(n - M)^i} = \theta + \sum_{i=1}^{\infty} \frac{a_i}{(N - 1)^i M^i},$$

and hence

$$E[J(\hat{\theta})] = \theta - \frac{a_2}{M^2 N(N - 1)} - \frac{a_3(2N - 1)}{M^3 N^2 (N - 1)^2} - \cdots$$

$$= \theta - \frac{a_2}{n(n - M)} - \frac{a_3(2n - M)}{n^2 (n - M)^2} - \cdots .$$

Thus the bias in $J(\hat{\theta})$ is of order n^{-2} while that of $\hat{\theta}$ is of order n^{-1} when $a_1 \neq 0$. Moreover if $a_2 = a_3 = \cdots = a_n = \cdots = 0$, as we have already seen from Theorem 2.1 or 2.2, $J(\hat{\theta})$ is unbiased. Some examples are now in order.

Example 3.1. Suppose that X_i $(i = 1,\ldots,n)$ are independent identically distributed (i.i.d.) as $N(\mu,\sigma^2)$. Further suppose that, whether due to reliance on the likelihood principle or only for purposes of illustration (as first done by Quenouille [34]), we are considering the use of

$$\hat{\theta} = \frac{1}{n} \sum_{i=1}^{n} (X_i - \overline{X})^2$$

$$= \frac{1}{n} \sum_{i=1}^{n} X_i^2 - \overline{X}^2 \quad ,$$

as an estimator of σ^2. Now it is clear that

$$E[\hat{\theta}] = \sigma^2 - \frac{1}{n} (\sigma^2) \quad .$$

But taking M = 1 we see that

$$\hat{\theta}^i = \frac{1}{n-1} \sum_{\substack{j=1 \\ j\neq i}}^{n} X_j^2 - \left(\frac{n\overline{X} - X_i}{n-1} \right)^2$$

$$= \frac{1}{n-1} \sum_{\substack{j=1 \\ j\neq i}}^{n} X_j^2 - \frac{1}{(n-1)^2} (n^2\overline{X}^2 - 2nX_i\overline{X} + X_i^2) \quad ,$$

and averaging over i we obtain

$$\hat{\theta}^{i} = \frac{1}{n} \sum_{i=1}^{n} \bar{\hat{\theta}}^{i}$$

$$= \frac{(n-1)^2 - 1}{n(n-1)^2} \sum_{i=1}^{n} X_i^2 + \frac{2n - n^2}{(n-1)^2} \bar{X}^2$$

$$= \frac{n^2 - 2n}{(n-1)^2} \hat{\theta} \quad .$$

Therefore,

$$E[\bar{\hat{\theta}}^{i}] = \frac{n^2 - 2n}{(n-1)^2} E[\hat{\theta}]$$

$$= \frac{n-2}{n-1} \sigma^2$$

$$= \sigma^2 - \left(\frac{1}{n-1} \right) (\sigma^2) \quad ,$$

and hence the choice of the parameter R(n) in the generalized jackknife in order that G be unbiased is given by

$$R(n) = \frac{\sigma^2/n}{\sigma^2/(n-1)} = \frac{n-1}{n} \quad .$$

Thus, in this example, the jackknife $J(\hat{\theta})$ and the generalized jackknife coincide. Continuing, we have

$$G(\hat{\theta}_1, \hat{\theta}_2) = J(\hat{\theta}) = n\hat{\theta} - (n-1)\bar{\theta}^i$$

$$= \sum_{i=1}^{n} X_i^2 - n\bar{X}^2 - \frac{n^2 - 2n}{(n-1)n} \sum_{i=1}^{n} X_i^2$$

$$+ \frac{n^2 - 2n}{n-1} \bar{X}^2$$

$$= \frac{n^2 - n - n^2 + 2n}{(n-1)n} \sum_{i=1}^{n} X_i^2$$

$$- \frac{n^2 - n - n^2 + 2n}{n-1} \bar{X}^2$$

$$= \frac{1}{n-1} \sum_{i=1}^{n} (X_i - \bar{X})^2 \quad .$$

This may be recognized as the unique minimum variance unbiased estimator of the variance of the normal distribution. In this case $J(\hat{\theta})$ is unbiased and is based upon the complete sufficient statistics alone and therefore we could not have expected any result other than that obtained. Of course it is not always the case that $J(\hat{\theta})$ will be a function of the sufficient statistics alone when $\hat{\theta}$ has that property, but, on the other hand, it is not uncommon.

Example 3.2. Suppose we have sampled the point binomial distribution with parameter p a total of n times. Then the number of "successes" define a random variable X, which has a binomial distribution, i.e., X has the density f defined by

$$f(x) = \binom{n}{x} p^x (1 - p)^{n-x}, \quad x = 0,1,\ldots,n \quad .$$

Further suppose that we desire an estimator of p^2, and because X/n is unbiased for p we are considering the estimator

$$\hat{\theta} = (X/n)^2 \quad .$$

It may be easily seen that

$$E[\hat{\theta}] = p^2 + \frac{1}{n}(p - p^2) \quad ,$$

and consequently we know that the ordinary jackknife will be unbiased. Note that the value of the estimator $\hat{\theta}^i$ will be one of two quantities, namely,

$$\left(\frac{x - 1}{n - 1}\right)^2 \quad \text{or} \quad \left(\frac{x}{n - 1}\right)^2 ,$$

depending upon whether the deleted trial is a success or a failure, respectively. Hence our second estimator $\hat{\theta}_2$, which is the average, $\bar{\hat{\theta}}^i$, of the n possible values of $\hat{\theta}^i$, is

$$\bar{\hat{\theta}}^i = \frac{1}{n} \left[X\left(\frac{X - 1}{n - 1}\right)^2 + (n - X)\left(\frac{X}{n - 1}\right)^2\right] ,$$

since there are X successes to be deleted and (n - X) failures. Simplifying this yields

$$\bar{\hat{\theta}}^i = \frac{(n - 2)X^2 + X}{n(n - 1)^2} ,$$

and thus

$$\begin{aligned} J(\hat{\theta}) &= n\hat{\theta} - (n - 1)\bar{\hat{\theta}}^i \\ &= \frac{X^2}{n} - \frac{(n - 2)X^2 + X}{n(n - 1)} \\ &= \frac{X(X - 1)}{n(n - 1)} . \end{aligned}$$

This estimator is unbiased for p^2 as anticipated and is also a function of the sufficient statistic alone.

In each of the above we have taken $R(n) = (n - 1)/n$. We will not make this assumption in general, although, henceforth we will assume $\hat{\theta}_2 = \overline{\hat{\theta}^{i}}$ and we will normally take $M = 1$. Some further examples which make use of different values of $R(n)$ will now be considered.

Example 3.3. Let X be uniformly distributed on the interval $(0,\theta)$, i.e., X has the density

$$f(x) = \begin{cases} \frac{1}{\theta} , & 0 < x < \theta \\ 0 , & \text{otherwise} . \end{cases}$$

A reasonable estimator for θ is

$$\hat{\theta} = \max \{X_1, X_2, \ldots, X_n\} = X_{(n)} .$$

It is well-known that $\hat{\theta}$ is biased and in fact since the expectation of the largest order statistic is $[n/(n + 1)]\theta$ we may write

$$E[\hat{\theta}] = \theta - \left(\frac{1}{n + 1} \right) \theta .$$

Even though it is obvious that $[(n + 1)/n]\hat{\theta}$ is unbiased, it is instructive to examine the jackknife statistic in this case. Note that

$$\hat{\theta}^{i} = \begin{cases} X_{(n)} , & \text{if } X_i \neq X_{(n)} \\ X_{(n-1)} , & \text{if } X_i = X_{(n)} \end{cases}$$

and hence

$$\bar{\hat{\theta}}^{i} = \frac{1}{n}\left[(n-1)X_{(n)} + X_{(n-1)}\right] .$$

Therefore Quenouille's method yields

$$\begin{aligned} J(\hat{\theta}) &= n\hat{\theta} - (n-1)\bar{\hat{\theta}}^{i} \\ &= nX_{(n)} - \left(\frac{(n-1)^2}{n}\right)X_{(n)} \\ &\qquad - \left(\frac{n-1}{n}\right)X_{(n-1)} \\ &= X_{(n)} + \left(\frac{n-1}{n}\right)\left[X_{(n)} - X_{(n-1)}\right] . \end{aligned}$$

The reduced bias may be observed by noting that

$$\begin{aligned} E[J(\hat{\theta})] &= \frac{n\theta}{n+1} + \left(\frac{n-1}{n}\right)\frac{\theta}{n+1} \\ &= \theta\left[1 - \frac{1}{n(n+1)}\right] . \end{aligned}$$

However, notice that in this problem the ratio

of the biases of $\hat{\theta}_1 = \hat{\theta}$ and $\hat{\theta}_2 = \bar{\hat{\theta}}^i$ is

$$\frac{E[\hat{\theta}_1 - \theta]}{E[\hat{\theta}_2 - \theta]} = \frac{\theta/(n+1)}{\theta/n} = \frac{n}{n+1} .$$

Consequently let us take the parameter

$$R(n) = \frac{n}{n+1}$$

and examine the statistic

$$G(\hat{\theta}_1, \hat{\theta}_2) = \frac{\hat{\theta}_1 - R(n)\hat{\theta}_2}{1 - R(n)} ,$$

which we now know to be unbiased. Then

$$\begin{aligned}\hat{\theta} &= (n+1)\hat{\theta}_1 - n\hat{\theta}_2 \\ &= (n+1)X_{(n)} - (n-1)X_{(n)} - X_{(n-1)} \\ &= X_{(n)} + \left[X_{(n)} - X_{(n-1)}\right] \\ &= 2X_{(n)} - X_{(n-1)} .\end{aligned}$$

It is interesting to note at this point that, if we take $\hat{\theta}_2 = X_{(n-1)}$, the bias of the second estimator is

$$E(\hat{\theta}_2) - \theta = \frac{n - 1}{n + 1} \theta - \theta$$

$$= \frac{2\theta}{n + 1} .$$

Hence in order that $G(\hat{\theta}_1, \hat{\theta}_2)$ be unbiased we must set

$$R(n) = \frac{\theta/(n + 1)}{2\theta/(n + 1)} = 1/2 .$$

Therefore

$$G(\hat{\theta}_1, \hat{\theta}_2) = \frac{X_{(n)} - 1/2\, X_{(n - 1)}}{1 - 1/2}$$

$$= 2X_{(n)} - X_{(n - 1)} .$$

Since this is identical to that obtained from the application of the generalized jackknife in the preceding paragraph, we are reminded of the possibility of using alternative procedures for selection of the second estimator. However, it should be mentioned again that no such technique has been put forth in the literature.

Let us now examine the mean square error (MSE) of this result relative to the MSE of the original estimator $\hat{\theta}_1$. We see that

$$\text{MSE}\ (\hat{\theta}) = E[(\hat{\theta} - \theta)^2]$$

$$= \frac{2\theta^2}{(n + 1)(n + 2)}$$

and, since $G(\hat{\theta}_1, \hat{\theta}_2)$ is unbiased,

$$\text{MSE}\ (G(\hat{\theta}_1, \hat{\theta}_2)) = \text{var}\ (G(\hat{\theta}_1, \hat{\theta}_2))$$

$$= \frac{2\theta^2}{(n + 1)(n + 2)} \quad .$$

Therefore, using the generalized jackknife we have completely eliminated the bias without increasing the mean square error. Before leaving this example let us examine the MSE of $J(\hat{\theta})$. Now

$$\text{MSE}\ (J(\hat{\theta})) = E[(J(\hat{\theta}) - \theta)^2]$$

$$= \frac{2\theta^2(n^2 - n + 1)}{n^2(n + 1)(n + 2)}, \quad n \geq 2.$$

Hence for all $n \geq 2$

$$\text{bias}\ (J(\hat{\theta})) < \text{bias}\ (\hat{\theta})$$

and

$$\text{MSE}\ (J(\hat{\theta})) < \text{MSE}\ (\hat{\theta}) \quad .$$

Thus it is clear that various choices of the parameter R(n) may serve to reduce the MSE as well as the bias.

<u>Example 3.4.</u> Suppose that the estimator $\hat{\theta}$ being considered has the property that

$$E[\hat{\theta} - \theta] = \frac{b}{n^2} .$$

It follows that

$$E[\overline{\hat{\theta}}^{i} - \theta] = \frac{b}{(n - 1)^2} .$$

The ordinary jackknife disregards this information in constructing the new estimator and we see that as a result

$$E[J(\hat{\theta}) - \theta] = n \frac{b}{n^2} - (n - 1) \frac{b}{(n - 1)^2}$$

$$= - \frac{b}{n(n - 1)} .$$

The bias of $J(\hat{\theta})$ has the opposite sign of the bias in $\hat{\theta}$ and is actually greater in magnitude. It should be clear from Theorem 2.1 that we need only take $R(n) = (n - 1)^2/n^2$ in the generalized jackknife to obtain an unbiased estimator for θ.

Example 3.5. Consider the problem of estimating the reliability function $r(x) = e^{-\lambda x}$. Let $t > 0$ be fixed and let the random variable N be defined by the number of failures occurring in the corresponding Poisson process in the test time t. The maximum likelihood estimator, $\hat{r}(x)$, for $r(x)$ is then given by

$$\hat{r}(x) = e^{-\frac{N}{t}x} .$$

Although the unique minimum variance unbiased estimator for $r(x)$ is known, the consideration of the reduction of the bias in $\hat{r}(x)$ by the jackknife method illustrates the potential usefulness of the parameter $R(n)$.

Now since

$$E[\hat{r}(x)] = \exp\,[-\lambda t(1 - e^{-x/t})] ,$$

we see that $\hat{r}(x)$ has a positive bias, and

$$\begin{aligned} E[\hat{r}(x)] &= \exp\,[-\lambda x + \lambda(x - t(1 - e^{-x/t}))] \\ &= e^{-\lambda x} \exp\,[\lambda(x - t(1 - e^{-x/t}))] \\ &= e^{-\lambda x} \exp\,[\lambda f(x,t)] \\ &= e^{-\lambda x}\,[1 + \lambda f(x,t) + \frac{\lambda^2}{2!} f^2(x,t) + \ldots] \\ &= r(x) + \sum_{k=1}^{\infty} e^{-\lambda x} \frac{\lambda^k}{k!} f^k(x,t) , \qquad (3.3) \end{aligned}$$

where $f(x,t) = x - t(1 - e^{-x/t})$. Note that the series expansion for the bias is composed of terms which are products of separate functions of the unknown parameter and of the observation time t (x is a known constant). We now take the approach used by Gaver and Hoel [20] to apply the jackknife in this problem.

Consider the time t to be split into n equal non-overlapping intervals of duration $t_i = t/n = \Delta t$ $(i = 1,\ldots,n)$. Let N_i be the number of failures in the ith interval. Hence, the N_i each have a Poisson distribution with parameter $\lambda t/n$, and N_i is independent of N_j for $i \neq j$. Now let $\hat{r}^{(i)}(x)$ denote the MLE of r obtained by deleting the ith interval from the collection of n intervals, e.g.,

$$\hat{r}^{(2)}(x) = \exp\left[-\left(\frac{N_1 + N_3 + \ldots + N_n}{(n-1)t/n}\right)x\right] .$$

Next define another estimator of $r(x)$, $\hat{\theta}_2$, to be the average of these n estimators, i.e.,

$$\hat{\theta}_2 = \frac{1}{n} \sum_{i=1}^{n} \hat{r}^{(i)}(x) .$$

By replacing t by $n\Delta t$ in (3.3) we see that

$$E[\hat{\theta}_2] = r(x) + \sum_{k=1}^{\infty} \left(e^{-\lambda x} \frac{\lambda^k}{k!}\right)(g^k(n;x,t)) ,$$

where

$$g(n;x,t) = f(x,(n-1)\Delta t) \quad .$$

Therefore letting $m = (n-1)/n$ and $h = x/t$ (the ratio of mission time to test time), we see that the appropriate $R(n)$ for combining $\hat{r}(x)$ and $\hat{\theta}_2$ to eliminate the first term of the expansion for the bias of $\hat{r}(x)$ is

$$R(n) = \frac{f(x,t)}{g(n;x,t)}$$

$$= \frac{x - t + te^{-h}}{x - mt + mte^{-h/m}}$$

$$= \frac{h - 1 + e^{-h}}{h - m + me^{-h/m}} \quad .$$

Note that $R(n)$ can also be written as

$$R(n) = \frac{(h^2/2) - (h^3/3!) + h^4/4!) - \ldots}{\dfrac{nh^2}{2(n-1)} - \dfrac{n^2h^3}{(n-1)^2 3!} + \dfrac{n^3h^4}{(n-1)^3 4!} - \ldots}$$

and hence for h sufficiently small, $R(n)$ is reasonably close to $(n-1)/n$. Consequently we see that the error incurred by using the jackknife in this problem may not be great in the cases where h is small.

In Example 3.4 we saw that an improper choice of R(n) can actually lead to an increase in bias in $\hat{\theta}$. Since a common practice when R(n) is not known is to assume $R(n) = (n - 1)/n$ it is interesting to know under what set of conditions this would lead to an actual increase in bias. A partial answer to this question is given below by Theorem 3.3. Let us first make a slight change in notation for convenience and also introduce some preliminary definitions and results. As we have already mentioned, for a given estimator $\hat{\theta}$ we will, unless specified to the contrary, always take $\hat{\theta}_1 = \hat{\theta}$ and $\hat{\theta}_2 = \bar{\hat{\theta}}^{i}$ in $G(\hat{\theta}_1,\hat{\theta}_2)$. When $\hat{\theta}_1$ and $\hat{\theta}_2$ are selected in this manner the notation $G(\hat{\theta}_1,\hat{\theta}_2)$ is needlessly cumbersome and we will adopt the shortened notation $G(\hat{\theta})$. Next we state those additional definitions which we shall require.

<u>DEFINITION 3.1</u>. Let $\hat{\theta}_1$ and $\hat{\theta}_2$ be estimators of θ defined on a sample of size n such that

$$B_1(n,\theta) = E[\hat{\theta}_1 - \theta] \neq 0$$

and

$$B_2(n,\theta) = E[\hat{\theta}_2 - \theta] \neq 0 \quad .$$

Then if

$$\left| \lim_{n\to\infty} \frac{B_1(n,\theta)}{B_2(n,\theta)} \right| = 1 \tag{3.4}$$

we will say that $\hat{\theta}_1$ and $\hat{\theta}_2$ are same order bias estimators of θ. This will be denoted $\hat{\theta}_1$ S.O.B.E. $\hat{\theta}_2$. If the absolute value of the limit in (3.4) is greater than zero and less than one we will say that $\hat{\theta}_1$ is a better same order bias estimator than $\hat{\theta}_2$ and denote this by $\hat{\theta}_1$ B.S.O.B.E. $\hat{\theta}_2$.

DEFINITION 3.2. If $\hat{\theta}_1$ and $\hat{\theta}_2$ are as in Definition 3.1 without the restriction $B_1(n,\theta) \neq 0$ and

$$\lim_{n\to\infty} \frac{B_1(n,\theta)}{B_2(n,\theta)} = 0 , \tag{3.5}$$

we will say that $\hat{\theta}_1$ is a lower order bias estimator that $\hat{\theta}_2$ and denote this $\hat{\theta}_1$ L.O.B.E. $\hat{\theta}_2$.

The following theorem will be useful in the remainder of this chapter. Its proof can be found in [8].

THEOREM 3.1. Let $\{a_n\}$ and $\{b_n\}$ be sequences of real numbers such that $\lim_{n\to\infty} a_n = 0$ and $\lim_{n\to\infty} b_n = 0$, where the latter is monotone. Then if

$$\lim_{n\to\infty} \frac{a_n - a_{n-1}}{b_n - b_{n-1}}$$

exists, finite or infinite, we have

$$\lim_{n\to\infty} \frac{a_n}{b_n} = \lim_{n\to\infty} \frac{\Delta a_n}{\Delta b_n} ,$$

where $\Delta a_n = a_n - a_{n-1}$ and $\Delta b_n = b_n - b_{n-1}$.

In order to characterize the value of the jackknife as a tool for bias reduction (which is what we are presently attempting to do) we need a general expression for the bias of $J(\hat{\theta})$ in terms of the bias of $\hat{\theta}$. This is accomplished by the next theorem.

THEOREM 3.2. Let $\hat{\theta}$ be an estimator for θ defined on the random sample $X_1, X_2, \ldots, X_n$. Further let

$$E[\hat{\theta} - \theta] = B(n,\theta)$$

and

$$\hat{\theta}_2 = \bar{\hat{\theta}}^i .$$

Then

$$E[J(\hat{\theta})] = \theta + B(n,\theta) + (n - 1)\Delta B(n,\theta) \qquad (3.6)$$

where

$$\Delta B(n,\theta) = B(n,\theta) - B(n - 1,\theta) .$$

Proof: Since $X_1, X_2, \ldots, X_n$ are independent

identically distributed random variables we have

$$E[\hat{\theta}_2 - \theta] = B(n-1,\theta) \quad .$$

But

$$J(\hat{\theta}) = n\hat{\theta} - (n-1)\hat{\theta}_2$$

$$= \hat{\theta} + (n-1)(\hat{\theta} - \hat{\theta}_2)$$

$$= \hat{\theta} + (n-1)[(\hat{\theta} - \theta) - (\hat{\theta}_2 - \theta)],$$

and hence

$$E[J(\hat{\theta})] = \theta + B(n,\theta) + (n-1)[B(n,\theta) - B(n-1,\theta)]$$

$$= \theta + B(n,\theta) + (n-1)\Delta B(n,\theta) \quad .$$

We are now in a position to establish criteria under which the jackknife is useful for bias reduction in an asymptotic sense.

THEOREM 3.3. If there exists a $p > 0$ such that

$$\lim_{n\to\infty} n^p B(n,\theta) = C(\theta) \neq 0 \quad \text{or} \quad \pm\infty,$$

and $\lim_{n\to\infty} n^{p+1}\Delta B(n,\theta)$ exists, then

(i) if $p = 1$, then $J(\hat{\theta})$ L.O.B.E. $\hat{\theta}$,

(ii) if $p < 2$ and $p \neq 1$, then $J(\hat{\theta})$ B.S.O.B.E. $\hat{\theta}$,

(iii) if $p = 2$, then $J(\hat{\theta})$ S.O.B.E. $\hat{\theta}$,

(iv) if $p > 2$, then $\hat{\theta}$ B.S.O.B.E. $J(\hat{\theta})$.

Proof: Since $C(\theta)$ is finite, $B(n,\theta) \to 0$ as $n \to \infty$. Moreover, n^{-p} converges monotonically to zero and the existence of $\lim_{n\to\infty} n^{p+1}\Delta B(n,\theta)$ implies

$\lim_{n\to\infty} [\Delta B(n,\theta)/\Delta(n^{-p})]$ exists. Thus by Theorem 3.1

$$C(\theta) = \lim_{n\to\infty} \frac{B(n,\theta)}{n^{-p}} = \lim_{n\to\infty} \frac{\Delta B(n,\theta)}{\Delta(n^{-p})} .$$

Now by Theorem 3.2

$$\left| \lim_{n\to\infty} \frac{E[J(\hat{\theta}) - \theta]}{E[\hat{\theta} - \theta]} \right|$$

$$= \left| \lim_{n\to\infty} \frac{B(n,\theta) + (n-1)\Delta B(n,\theta)}{B(n,\theta)} \right|$$

$$= \left| 1 + \lim_{n\to\infty} \frac{(n-1)\Delta B(n,\theta)}{B(n,\theta)} \right|$$

$$= \left| 1 + \lim_{n\to\infty} \frac{(n-1)^{p+1}\Delta(n^{-p})[\Delta B(n,\theta)/\Delta(n^{-p})]}{(n-1)^p B(n,\theta)} \right|$$

$$= \left| 1 + \lim_{n\to\infty} \frac{(n - 1)^{p+1}\Delta(n^{-p})C(\theta)}{C(\Theta)} \right| .$$

But by simple algebra

$$\lim_{n\to\infty} (n - 1)^{p+1}\Delta(n^{-p}) = -p .$$

Consequently

$$\left| \lim_{n\to\infty} \frac{E[J(\hat{\theta}) - \theta]}{E[\hat{\theta} - \theta]} \right| = \left| 1 - p \right| , \qquad (3.7)$$

and (i) through (iv) follow.

<u>Example 3.6.</u>

(a) To illustrate Theorem 3.3 let us reconsider Example 3.1 in which we had

$$\hat{\theta} = \frac{1}{n} \sum_{i=1}^{n} (X_i - \overline{X})^2$$

as an estimator of σ^2. Consequently

$$B(n,\sigma^2) = -\frac{1}{n}\ \sigma^2 ,$$

so that p = 1, i.e.,

$$\lim_{n\to\infty} n^1 B(n,\sigma^2) = -\sigma^2 = C(\sigma^2) .$$

Moreover it is easy to see that $\lim_{n\to\infty} n^2 \Delta B(n,\theta)$ exists. Hence from part (i) of Theorem 3.3 we are assured that $J[\hat{\theta}]$ L.O.B.E. $\hat{\theta}$; and indeed, since $J[\hat{\theta}]$ is unbiased,

$$\lim_{n\to\infty} \frac{\text{bias }[J(\hat{\theta})]}{\text{bias }[\hat{\theta}]} = \lim_{n\to\infty} \frac{0}{-\frac{1}{n}\sigma^2} = 0 ,$$

which, from Definition 3.2, shows that $J(\hat{\theta})$ is a lower order bias estimator than $\hat{\theta}$.

(b) Next recall that in Example 3.3 we were examining $\hat{\theta} = X_{(n)}$ (the largest order statistic) as an estimator of the truncation point θ of the uniform distribution. Here we have

$$B(n,\theta) = -\frac{1}{n+1}\theta$$

and again $p = 1$. Hence from Theorem 3.3 (i) we have once more $J(\hat{\theta})$ L.O.B.E. $\hat{\theta}$ although in this instance $J(\hat{\theta})$ is not unbiased. In particular

$$\lim_{n\to\infty} \frac{\text{bias }[J(\hat{\theta})]}{\text{bias }[\hat{\theta}]} = \lim_{n\to\infty} \left[-\frac{1}{n(n+1)}\theta \Big/ -\frac{1}{n+1}\theta\right]$$

$$= \lim_{n\to\infty} \frac{1}{n} = 0 .$$

(c) Now let us suppose the distribution of our estimator is such that

$$E[\hat{\theta}] - \theta = B(n,\theta)$$

$$= b(\theta)/\sqrt{n} \quad .$$

In this event p = 1/2 and part (iii) of Theorem 3.3 is relevant. It is easy to see that

$$E[J(\hat{\theta})] = n\ E[\hat{\theta}] - (n - 1)\ E[\bar{\hat{\theta}}^{i}]$$

$$= \theta + (\sqrt{n} - \sqrt{n - 1})\,b(\theta) \quad ,$$

and therefore

$$\frac{\text{bias } [J(\hat{\theta})]}{\text{bias } [\hat{\theta}]} = \sqrt{n^2} - \sqrt{n(n - 1)}$$

$$= n - n(1 - \tfrac{1}{n})^{1/2}$$

$$= n - n\left(1 - \frac{1}{2n} - \frac{1}{8n^2} - \frac{1}{16n^3} - \cdots\right)$$

$$= \frac{1}{2} + \frac{1}{8n} + \frac{1}{16n^2} + \cdots \quad .$$

Therefore

$$\lim_{n\to\infty} \frac{\text{bias } [J(\hat{\theta})]}{\text{bias } [\hat{\theta}]} = \frac{1}{2}$$

indicating that $J(\hat{\theta})$ B.S.O.B.E. $\hat{\theta}$ as promised.

(d) Example 3.4 provides an illustration of part (iii) of the theorem since in that example

$$B(n,\theta) = \frac{b}{n^2} \quad .$$

Thus we know that $J(\hat{\theta})$ and $\hat{\theta}$ are same order bias estimators and indeed

$$\frac{\text{bias } [J(\hat{\theta})]}{\text{bias } (\hat{\theta})} = \frac{-b/n(n - 1)}{b/n^2} = - \frac{n}{n - 1}$$

so that

$$\lim_{n\to\infty} \frac{\text{bias } [J(\hat{\theta})]}{\text{bias } [\hat{\theta}]} = 1 \quad .$$

Recall that in Example 3.4 the magnitude of the bias was actually greater after jackknifing than before. This however is not at all out of line with the notion that $J(\hat{\theta})$ S.O.B.E. $\hat{\theta}$.

(e) Finally if $\hat{\theta}$ were such that

$$B(n,\theta) = \frac{a}{n^3} \quad ,$$

then the bias of $J(\hat{\theta})$ would be

$$E[J(\hat{\theta})] - \theta = n \frac{a}{n^3} - (n - 1) \frac{a}{(n - 1)^3}$$

$$= a \left[\frac{- 2n + 1}{n^2 (n - 1)^2} \right] .$$

Consequently

$$\frac{\text{bias } [J(\hat{\theta})]}{B(n,\theta)} = \frac{(-2n + 1)n}{(n - 1)^2}$$

$$= - \frac{2 - (1/n)}{1 - (2/n) + (1/n^2)}$$

and

$$\left| \lim_{n\to\infty} \frac{\text{bias } [J(\hat{\theta})]}{B(n,\theta)} \right| = 2 \quad .$$

Consequently $\hat{\theta}$ B.S.O.B.E. $J(\hat{\theta})$ which we could have seen from part (iv) since here $p = 3$.

We have for the present, focused special attention on $G(\hat{\theta})$ when $R(n) = (n - 1)/n$. We have done so for two reasons: (1) this is the most frequent situation in practice and (2) this is the most extensively studied case. It would not, however, be proper to leave the impression that similar results cannot be obtained in other cases which in some instances may be even more interesting. This is exemplified by the following theorem where it is shown that when one correctly assumes the order of the bias in $\hat{\theta}$ (whether it is $0(n^{-1})$ is immaterial) then practically speaking one can always produce an L.O.B.E.

<u>THEOREM 3.4</u>. Let $B(n,\theta) \neq 0$ and $p > 0$ be as in Theorem 3.3. Then if

$$R(n) = \left(\frac{n-1}{n}\right)^m, \quad m > 0,$$

we have

(i) if $p = m$, $G(\hat{\theta})$ L.O.B.E. $\hat{\theta}$,

(ii) if $0 < p < 2m$ and $p \neq m$, $G(\hat{\theta})$ B.S.O.B.E. $\hat{\theta}$,

(iii) if $p = 2m$, $G(\hat{\theta})$ S.O.B.E. $\hat{\theta}$,

(iv) if $p > 2m$, $\hat{\theta}$ B.S.O.B.E. $G(\hat{\theta})$,

(v) if $p = m \neq 1$, $G(\hat{\theta})$ L.O.B.E. $J(\hat{\theta})$,

(vi) if $p > 1$ and $1 < m < p(2-p)^{-1}$ or $p < 1$ and $p(2-p)^{-1} < m < 1$, $G(\hat{\theta})$ B.S.O.B.E. $J(\hat{\theta})$.

(vii) if $p = m = 1$, $G(\hat{\theta})$ S.O.B.E. $J(\hat{\theta})$,

(viii) if $p > 1$ and $m < 1$ or $p < 1$ and $m > 1$, $J(\hat{\theta})$ B.S.O.B.E. $G(\hat{\theta})$,

(ix) if $p = 1$ and $m \neq 1$, $J(\hat{\theta})$ L.O.B.E. $G(\hat{\theta})$.

Before proceeding with the proof of Theorem 3.4 let us make some comments concerning it, in order to shed some light on its meaning. Roughly speaking, the theorem gives criteria under which $G(\hat{\theta})$ would be expected to be useful for bias reduction, as well as conditions for which $G(\hat{\theta})$ would be expected to be better than $J(\hat{\theta})$ for this purpose. Moreover the theorem also establishes the degree of accuracy to which one must know the order of the bias in $\hat{\theta}$ in order for $G(\hat{\theta})$ to be effective as a bias reduction tool. Thus for example if one <u>postulates</u> $B(n,\theta) = 0(n^{-2/3})$ one can expect good results from $G(\hat{\theta})$ relative to $\hat{\theta}$ with $m = 2/3$ so long as the true bias satisfies the relation $B(n,\theta) = 0(n^{-p})$ for some $p < 4/3$, i.e., provided the bias as a function of $1/n$ is of lower order than 4/3. Note from (ii) that by taking m large one can essentially guarantee himself that $G(\hat{\theta})$ will not increase the bias. However, since $\lim_{n\to\infty} R(n) = 0$, one also sees that the larger he takes m the less he will alter $\hat{\theta}$ and hence any gains he obtains in bias reduction would probably occur only for large samples where the bias is rather insignificant anyway. Moreover the range of m for which $G(\hat{\theta})$ is better than $J(\hat{\theta})$ is a bounded function of p and hence rather than pick m large one is better off to analyze the bias, examine the results of Theorem 3.4, and, based on the estimated range of the order of the bias, select the appropriate estimator, $\hat{\theta}$, $J(\hat{\theta})$, or $G(\hat{\theta})$. We now proceed with the proof of Theorem 3.4.

Proof:

$$\frac{E[G(\hat{\theta}) - \theta]}{E[\hat{\theta} - \theta]} = \frac{B(n,\theta) - R(n)B(n - 1,\theta)}{(1 - R(n))B(n,\theta)}$$

$$= 1 + \frac{R(n)}{1 - R(n)} \frac{\Delta B(n,\theta)}{B(n,\theta)}$$

$$= 1 + \frac{R(n)}{1 - R(n)} \cdot \frac{1}{n - 1} [(n - 1)\frac{\Delta B(n,\theta)}{B(n,\theta)}].$$

But from the proof of Theorem 3.3

$$\lim_{n\to\infty} (n - 1) \frac{\Delta B(n,\theta)}{B(n,\theta)} = -p \quad .$$

However,

$$\frac{R(n)}{1 - R(n)} \frac{1}{n - 1} = \frac{(n - 1)^m}{n^m - (n - 1)^m} \frac{1}{n - 1}$$

and hence

$$\lim_{n\to\infty} \frac{R(n)}{1 - R(n)} \cdot \frac{1}{n - 1} = \frac{1}{m} \quad .$$

It follows then that

$$\lim_{n\to\infty} \frac{E[G(\hat{\theta}) - \theta]}{E[\hat{\theta} - \theta]} = 1 - \frac{p}{m} . \tag{3.8}$$

From (3.8), (i) follows at once. To see (ii) note that if $p \neq m$, then

$$\left| 1 - \frac{p}{m} \right| < 1$$

if and only if $0 < p < 2m$. The results (iii) and (iv) also follow easily from (3.8).

To show the remainder of the theorem note that

$$\frac{E[G(\hat{\theta}) - \theta]}{E[J(\hat{\theta}) - \theta]} = [1 + \frac{R(n)}{1 - R(n)} \frac{\Delta B(n,\theta)}{B(n,\theta)}] \bigg/ \left(1 + (n - 1) \frac{\Delta B(n,\theta)}{B(n,\theta)}\right) .$$

Thus if $p \neq 1$

$$\left| \lim_{n\to\infty} \frac{E[G(\hat{\theta}) - \theta]}{E[J(\hat{\theta}) - \theta]} \right| = \left| \frac{1 - p/m}{1 - p} \right| . \tag{3.9}$$

Examination of (3.9) establishes (v), (vi) and (viii) and consideration of its reciprocal provides the proof of (ix) where $p = 1$. Part (vii) where $p = m = 1$ follows trivially since then $G(\hat{\theta}) = J(\hat{\theta})$ and the proof is complete.

4. The Higher Order Generalized Jackknife

Consider once more the estimator

$$G(\hat{\theta}_1, \hat{\theta}_2) = \frac{\hat{\theta}_1 - R(n)\hat{\theta}_2}{1 - R(n)} , \tag{4.1}$$

and assume

$$R(n) = \frac{a_{11}(n)}{a_{12}(n)} ,$$

which would arise if

$$B_k(n, \theta) = a_{1k}(n) b_1(\theta), \quad k = 1,2 .$$

Then suppressing the n we can write

$$G(\hat{\theta}_1, \hat{\theta}_2) = \frac{\begin{vmatrix} \hat{\theta}_1 & \hat{\theta}_2 \\ a_{11} & a_{12} \end{vmatrix}}{\begin{vmatrix} 1 & 1 \\ a_{11} & a_{12} \end{vmatrix}} , \tag{4.2}$$

which suggests the following extension.

DEFINITION 4.1. Let $\hat{\theta}_1, \hat{\theta}_2, \hat{\theta}_3, \ldots, \hat{\theta}_{k+1}$ be k + 1 estimators for θ defined on the random sample $X_1, X_2, \ldots, X_n$. Further let a_{ij}, i = 1,2,...,k, j = 1,...,k+1 be real numbers such that

$$\begin{vmatrix} 1 & 1 & \cdots & 1 \\ a_{11} & a_{12} & \cdots & a_{1,k+1} \\ \vdots & \vdots & & \vdots \\ a_{k1} & a_{k2} & \cdots & a_{k,k+1} \end{vmatrix} \neq 0 . \qquad (4.3)$$

Then we define $G(\hat{\theta}_1, \hat{\theta}_2, \ldots, \hat{\theta}_{k+1})$ by

$$G(\hat{\theta}_1, \hat{\theta}_2, \ldots, \hat{\theta}_{k+1}) = \frac{\begin{vmatrix} \hat{\theta}_1 & \hat{\theta}_2 & \cdots & \hat{\theta}_{k+1} \\ a_{11} & a_{12} & \cdots & a_{1,k+1} \\ \vdots & \vdots & & \vdots \\ a_{k1} & a_{k2} & \cdots & a_{k,k+1} \end{vmatrix}}{\begin{vmatrix} 1 & 1 & \cdots & 1 \\ a_{11} & a_{12} & \cdots & a_{1,k+1} \\ \vdots & \vdots & & \vdots \\ a_{k1} & a_{k2} & \cdots & a_{k,k+1} \end{vmatrix}} . \qquad (4.4)$$

When the bias in the estimator $\hat{\theta}_1, \hat{\theta}_2, \ldots, \hat{\theta}_{k+1}$ can be written as the product of a function of n and a function of θ we have the following generalization of Theorem 2.1.

THEOREM 4.1. If

$$E[\hat{\theta}_j] - \theta = \sum_{i=1}^{\infty} f_{ij}(n)b_i(\theta), \quad j = 1,2,\ldots,k+1 ,$$

and if equation (4.3) is satisfied with

$$a_{ij} = f_{ij}(n) \quad ,$$

then

$$E[G(\hat{\theta}_1,\hat{\theta}_2,\ldots,\hat{\theta}_{k+1})] = \theta + B_G(n,\theta) \quad ,$$

where

$$B_G(n,\theta) = \frac{\begin{vmatrix} B_1 & B_2 & \cdots & B_{k+1} \\ f_{11}(n) & f_{12}(n) & \cdots & f_{1,k+1}(n) \\ \vdots & & & \vdots \\ f_{k1}(n) & & \cdots & f_{k,k+1}(n) \end{vmatrix}}{\begin{vmatrix} 1 & 1 & \cdots & 1 \\ f_{11}(n) & & \cdots & f_{1,k+1}(n) \\ \vdots & & & \vdots \\ f_{k1}(n) & & \cdots & f_{k,k+1}(n) \end{vmatrix}}$$

and

$$B_j = \sum_{i=k+1}^{\infty} f_{ij}(n) b_i(\theta), \quad j = 1,2,\ldots,k+1 .$$

Proof: The result immediately follows by noting that substituting $a_{ij} = f_{ij}(n)$ in (4.4) yields

$$E[G(\hat{\theta}_1, \hat{\theta}_2, \ldots, \hat{\theta}_{k+1})] =$$

$$\theta + \frac{\begin{vmatrix} \sum_{i=1}^{\infty} f_{i1}(n) b_i(\theta) & \cdots & \sum_{i=1}^{\infty} f_{ik+1}(n) b_i(\theta) \\ f_{11}(n) & \cdots & f_{1k+1}(n) \\ \vdots & & \vdots \\ f_{k1}(n) & \cdots & f_{k,k+1}(n) \end{vmatrix}}{\begin{vmatrix} 1 & 1 & \cdots & 1 \\ f_{11}(n) & & \cdots & f_{1k+1}(n) \\ \vdots & & & \vdots \\ f_{k1}(n) & & \cdots & f_{k,k+1}(n) \end{vmatrix}}$$

and hence the theorem follows by elementary row operations.

<u>COROLLARY 4.1.</u> If

$$E[\hat{\theta}_j] = \theta + \sum_{i=1}^{k} f_{ij}(n) b_i(\theta), \quad j = 1,2,\ldots,k+1 ,$$

then

$$E[G(\hat{\theta}_1, \hat{\theta}_2, \ldots, \hat{\theta}_{k+1})] = \theta \quad .$$

That is, if the $f_{ij}(n)$ are known and the a_{ij} are defined by

$$a_{ij} = f_{ij}(n) \quad ,$$

then $G(\hat{\theta}_1, \hat{\theta}_2, \ldots, \hat{\theta}_{k+1})$ is an unbiased estimator of θ.

Henceforth we will assume the bias in $\hat{\theta}_j$ is of the form

$$\sum_{i=1}^{\infty} f_{ij}(n) b_i(\theta)$$

and take

$$a_{ij} = f_{ij}(n), \; i = 1,2,\ldots,k, \quad j = 1,2,\ldots,k+1.$$

When the $f_{ij}(n)$ in Theorem 4.1 have the particular form $f_{ij}(n) = 1/(n - j + 1)^i$ the result of that theorem can be expressed quite nicely. This is the subject of the next theorem. The proof is given in [44].

<u>THEOREM 4.2</u>. If the bias in $\hat{\theta}_j$, $j = 1,2,\ldots,k + 1$ is as in Theorem 4.1 and, for all i and j, $f_{ij}(n) = 1/(n - j + 1)^i$, then $E[G(\hat{\theta}_1, \hat{\theta}_2, \ldots, \hat{\theta}_{k+1}) - \theta] = 0(n^{-k-1})$. That is, the biases in $\hat{\theta}_1, \hat{\theta}_2, \ldots, \hat{\theta}_{k+1}$ are of order n^{-1} while the bias in $G(\hat{\theta}_1, \hat{\theta}_2, \ldots, \hat{\theta}_{k+1})$ is of order n^{-k-1}. Further if

$$E[\hat{\theta}_j] = \theta + \sum_{i=1}^{k} \frac{b_i(\theta)}{(n - j + 1)^i}$$

then $G(\hat{\theta}_1, \hat{\theta}_2, \ldots, \hat{\theta}_{k+1})$ is unbiased.

In the above theorems no mention has been made of how the k estimators $\hat{\theta}_2, \hat{\theta}_3, \ldots, \hat{\theta}_{k+1}$ should be selected for a given $\hat{\theta}_1$. One method for selecting these estimators is to simply extend the method of Quenouille as follows. Let $k < n$ and let $\hat{\theta}_1^{i_2}, \hat{\theta}_1^{i_2 i_3}, \ldots, \hat{\theta}_1^{i_2 \ldots i_{k+1}}$ be the estimators resulting by by restricting $\hat{\theta}_1$ to the samples obtained by deleting at random 1,2,...,k of the observations, respectively. Then define

$$\hat{\theta}_j = \overline{\hat{\theta}_1^{i_2, \ldots, i_j}} \quad , \quad j = 2,3,\ldots,k + 1 \ , \qquad (4.5)$$

the average over the $\binom{n}{j-1}$ resulting statistics.

Note that when $\hat{\theta}_j$ is selected in this way,

$$E[\hat{\theta}_j - \theta] = B(n - j + 1, \theta) \quad ,$$

and hence if $\hat{\theta}_{k+1}$ is defined and $\hat{\theta}_1$ is unbiased, $G(\hat{\theta}_1, \hat{\theta}_2, \ldots, \hat{\theta}_{k+1})$ is also unbiased. When (4.5) holds and the bias in $\hat{\theta}_1$ is a power series in $1/n$, i.e.,

$$E[\hat{\theta}_1 - \theta] = \frac{a_1}{n} + \frac{a_2}{n^2} + \cdots , \tag{4.6}$$

we have

$$E[\hat{\theta}_j - \theta] = \frac{a_1}{n - j + 1} + \frac{a_2}{(n - j + 1)^2} + \cdots .$$

Hence when (4.5) and (4.6) hold

$$G(\hat{\theta}_1, \hat{\theta}_2, \ldots, \hat{\theta}_{k+1})$$

$$= \frac{\begin{vmatrix} \hat{\theta}_1 & \hat{\theta}_2 & \cdots & \hat{\theta}_{k+1} \\ \frac{1}{n} & \frac{1}{n-1} & \cdots & \frac{1}{n-k} \\ \vdots & \vdots & & \vdots \\ \frac{1}{n^k} & \frac{1}{(n-1)^k} & \cdots & \frac{1}{(n-k)^k} \end{vmatrix}}{\begin{vmatrix} 1 & 1 & \cdots & 1 \\ \frac{1}{n} & \frac{1}{n-1} & \cdots & \frac{1}{n-k} \\ \vdots & \vdots & & \vdots \\ \frac{1}{n^k} & \frac{1}{(n-1)^k} & \cdots & \frac{1}{(n-k)^k} \end{vmatrix}} \tag{4.7}$$

and by Theorem 4.2 $E[G(\hat{\theta}_1,\hat{\theta}_2,\ldots,\hat{\theta}_{k+1}) - \theta] = O(n^{-k-1})$ if the $a_i \neq 0$, and $G(\hat{\theta}_1,\hat{\theta}_2,\ldots,\hat{\theta}_{k+1})$ is unbiased if $a_{k+1} = a_{k+2} = \ldots = 0$. When referring to the particular form of $G(\hat{\theta}_1,\hat{\theta}_2,\ldots,\hat{\theta}_{k+1})$ given by (4.7) we will henceforth take $\hat{\theta} = \hat{\theta}_1$ and use the notation $J^{(k)}(\hat{\theta})$. Moreover, given a particular estimator $\hat{\theta}$, we will always select the remaining k estimators in the manner described by equation (4.5). Consequently we will now adopt the more manageable notation $G^{(k)}(\hat{\theta})$ in place of $G(\hat{\theta}_1,\hat{\theta}_2,\ldots,\hat{\theta}_{k+1})$. The following relationships should now be clear:

$$G(\hat{\theta}_1,\hat{\theta}_2,\ldots,\hat{\theta}_{k+1}) = G^{(k)}(\hat{\theta})$$

when $\hat{\theta}_1 = \hat{\theta}$ and the $\hat{\theta}_j$ $(j = 2,\ldots,k + 1)$ are given by (4.5), and

$$G(\hat{\theta}_1,\hat{\theta}_2) = G^{(1)}(\hat{\theta}) = G(\hat{\theta}) \quad .$$

Furthermore

$$G(\hat{\theta}) = J(\hat{\theta}) \quad , \quad \text{if} \quad R(n) = (n - 1)/n \quad ,$$

and $G^{(k)}(\hat{\theta}) = J^{(k)}(\hat{\theta})$ when $f_{ij}(n) = [1/(n - j + 1)^i]$.

<u>Example 4.1.</u> Suppose $\hat{\theta}$ is an estimator for θ such that

$$B(n,\theta) = \frac{b_1(\theta)}{n} + \frac{b_2(\theta)}{n^2} \quad ,$$

and consider $J^{(2)}(\hat{\theta})$. By Theorem 4.2 $J^{(2)}(\hat{\theta})$ is unbiased. Note however that not only is $J(\hat{\theta})$ biased, but if $b_1(\theta) = 0$ (see Example 3.4) the magnitude of the bias in $J(\hat{\theta})$ is actually greater than that of $\hat{\theta}$.

Example 4.2. To illustrate the higher order jackknife we shall consider the general class of problems suggested by Example 3.3. In many practical applications of statistical theory the random variables which enter the mathematical model are restricted to a finite range; even though the assumed distribution, which is satisfactory in all other aspects, may ordinarily be defined upon random variables whose range is unbounded. In these instances the estimation of the truncation point may be of definite practical importance.

Suppose an estimate of a truncation point θ is desired and we have a random sample of size n from the distribution F(x) which has been truncated to $x \leq \theta$, i.e.,

$$X \sim F_\theta(x) = F(x)/F(\theta) \quad .$$

A natural choice for the estimator is the largest of the n observations,

$$\hat{\theta} = \max \{X_1, X_2, \ldots, X_n\} = X_{(n)} \quad .$$

Following Robson and Whitlock [41] the bias in $\hat{\theta}$ may be computed if we make the probability transformation

$$Y = F_\theta(X_{(n)}) = F(X_{(n)})/F(\theta) \quad ,$$

so that

$$X_{(n)} = H_{F_\theta}(Y) \quad .$$

Second, the function $H_{F_\theta}(Y)$ may be expanded in a Taylor series about the point Y = 1, to yield (omitting the subscript for H)

$$X_{(n)} = \theta + (Y - 1)H'(1) + \frac{(Y - 1)^2}{2!} H''(1) + \frac{(Y - 1)^3}{3!} H'''(1) + \ldots , \qquad (4.8)$$

where

$$H(1) = \theta \quad ,$$

$$H'(1) = \frac{F(\theta)}{F'(\theta)} \quad ,$$

$$H''(1) = -\left[\frac{F(\theta)}{F'(\theta)}\right]^2 \frac{F''(\theta)}{F'(\theta)} \quad ,$$

$$H'''(1) = \left[\frac{F(\theta)}{F'(\theta)}\right]^3 \left\{3\left[\frac{F''(\theta)}{F'(\theta)}\right]^2 - \frac{F'''(\theta)}{F'(\theta)}\right\} \quad ,$$

and so forth.

Since (1 - Y) is distributed as Y_1, the smallest of a random sample of size n from the uniform distribution over (0,1), we have

$$E(Y - 1)^k = (-1)^k E(Y_1^k) = (-1)^k \frac{k!n!}{(n + k)!} .$$

Therefore

$$E(X_{(n)}) = \theta - \left(\frac{1}{n + 1}\right) H'(1) + \frac{1}{(n + 1)(n + 2)} H''(1)$$

$$- \frac{n!}{(n + 3)!} H'''(1) + \ldots \quad . \tag{4.9}$$

Recall that in Example 3.3 we obtained

$$\bar{\bar{\theta}}^{i} = \frac{1}{n} \left[(n - 1)X_{(n)} + X_{(n-1)}\right] ,$$

which leads to

$$G(\hat{\theta}) = 2X_{(n)} - X_{(n-1)} \quad .$$

The use of this estimator in the class of distributions under consideration leads to a bias given by

$$E[G(\hat{\theta}) - \theta] = -\frac{1}{(n+1)(n+2)} H''(1) + \frac{2n!}{(n+3)!} H'''(1) - \frac{3n!}{(n+4)!} H^{(4)}(1) + \cdots .$$

In Example 3.3 X had a uniform distribution over $(0,\theta)$ or

$$F_\theta(x) = \frac{x}{\theta}, \quad 0 < x < \theta ,$$

and consequently $H'_{F_\theta}(1) = \theta$ is the only non-vanishing derivative in equation (4.8). Hence $G(\hat{\theta})$ is unbiased as we have already shown.

Now if we evaluate $\hat{\theta}$ over all possible sub-samples of size two and average we obtain

$$\hat{\theta}_3 = \bar{\hat{\theta}}^{ij} = \frac{2}{n(n-1)} \left[\frac{(n-1)(n-2)}{2} X_{(n)} + (n-2) X_{(n-1)} + X_{(n-2)} \right] .$$

In the notation of this section we set

$$a_{11} = f_{11}(n) = \frac{1}{n+1} ,$$

$$a_{21} = f_{21}(n) = \frac{1}{(n + 1)(n + 2)}$$

and thus

$$a_{12} = \frac{1}{n} \quad , \qquad a_{22} = \frac{1}{n(n + 1)} \quad ,$$

$$a_{13} = \frac{1}{n - 1} , \quad a_{23} = \frac{1}{n(n - 1)} \quad .$$

Hence

$$G^{(2)}(\hat{\theta}) =$$

$$\begin{vmatrix} \hat{\theta}_1 & \hat{\theta}_2 & \hat{\theta}_3 \\ \frac{1}{n + 1} & \frac{1}{n} & \frac{1}{n - 1} \\ \frac{1}{(n + 1)(n + 2)} & \frac{1}{n(n + 1)} & \frac{1}{n(n - 1)} \end{vmatrix} \cdot \left[\sum_{j=1}^{3} c_2^j\right]^{-1}$$

where

$$c_2^1 = \left[\frac{1}{n^2(n - 1)} - \frac{1}{n(n - 1)(n + 1)}\right] = \frac{1}{n^2(n - 1)(n + 1)} \quad ,$$

$$c_2^2 = \left[\frac{1}{(n-1)(n+1)(n+2)} - \frac{1}{n(n-1)(n+1)} \right]$$

$$= \frac{-2}{n(n-1)(n+1)(n+2)}$$

and

$$c_2^3 = \left[\frac{1}{n(n+1)^2} - \frac{1}{n(n+1)(n+2)} \right]$$

$$= \frac{1}{n(n+1)^2(n+2)} \quad .$$

Now

$$\sum_{j=1}^{3} c_2^j = \frac{2}{n^2(n+1)^2(n-1)(n+2)} \quad ,$$

and therefore

$$G^{(2)}(\hat{\theta}) = \frac{1}{2}\left[(n+1)(n+2)\hat{\theta}_1 - 2n(n+1)\hat{\theta}_2 + n(n-1)\hat{\theta}_3\right]$$

$$= \frac{1}{2}\left[(n+1)(n+2) - 2(n+1)(n-1) + (n-1)(n-2)\right] X_{(n)}$$

$$+ \frac{1}{2}[2(n-2) - 2(n+1)]X_{(n-1)} + X_{(n-2)}$$

$$= 3X_{(n)} - 3X_{(n-1)} + X_{(n-2)} \quad .$$

This is the same estimator which Robson and Whitlock [41] derive by adhering to the proper principles for reapplication of a bias reduction scheme. The bias of $G^{(2)}(\hat{\theta})$ is

$$E[G^{(2)}(\hat{\theta}) - \theta] = -\frac{n!}{(n+3)!}H'''(1) + \frac{3n!}{(n+4)!}H^{(4)}(1) - \cdots$$

which is clearly $O(n^{-3})$.

It may be recalled that in Example 3.3 $G(\hat{\theta})$ was shown to have the same mean square error as $\hat{\theta}$ when the underlying distribution is uniform. One major question that will occur to many readers is whether or not these higher order transformations can possibily avoid a pronounced increase in MSE. In the present example the comparison of mean square errors for G and $G^{(2)}$ is unfair to the second order estimator since G is unbiased and the higher order transformation is not required. Indeed it can be shown that the inclusion of $X_{(n-2)}$ in $G^{(2)}$ results in

$$\mathrm{MSE}\left[\; G^{(2)}\;(\hat{\theta}_1)\;\right] \;=\; 3\mathrm{MSE}\left[\; G\;(\hat{\theta}_1)\;\right] .$$

Such severe penalties in variance will not always be the price for elimination of additional terms in the bias expansion as we shall see in some of the examples which follow here and in Chapter II.

<u>Example 4.3.</u> Consider the geometric distribution as the simple model for representing the number of independent trials required to obtain an occurrence of an event, which we shall refer to as a "failure." The density of X, the number of "successes" prior to the failure, is given by

$$f_X(x) = p^x q \;, \qquad x = 0, 1, 2, \ldots \quad .$$

If this distribution is sampled n times and the

observations on X are summed and denoted by Y, i.e.,

$$Y = \sum_{i=1}^{n} X_i \quad ,$$

then Y has the negative binomial distribution with density

$$f_Y(y) = \binom{n + y - 1}{y} p^y q^n, \quad y = 0,1,2,\ldots \quad .$$

Frequently this arises as the probability density of the total number of trials (n + y) which will be required to obtain n failures.

It is easily shown that

$$E[X] = p/q \quad \text{and var } [X] = p/q^2 \quad .$$

Further, if we have the aforementioned random sample of size n, then the maximum likelihood estimate of p is given by

$$\hat{p} = \frac{Y}{n + Y} \quad ,$$

which may be modified to yield the minimum variance unbiased estimate based on the complete sufficient statistic Y; viz.,

$$\hat{\hat{p}} = \frac{Y}{n + Y - 1} \quad .$$

Now suppose we desire an estimator for the mean of the X population and because of the invariance property of MLE we use the corresponding one-to-one function of $\hat{p}$, i.e., we take

$$\hat{\mu}_X = \frac{\hat{p}}{\hat{q}} = \frac{Y}{n + Y} \bigg/ \frac{n}{n + Y} = \frac{Y}{n} .$$

We may express the expectation of this latter estimator as

$$E[\hat{\hat{\mu}}_X] = \frac{p}{q} + \left(\frac{1}{n - 1}\right)\frac{p}{q} ,$$

which would dictate the use of

$$R(n) = \frac{n - 2}{n - 1}$$

in the generalized jackknife. It is easy to see that doing so in this example yields

$$G(\hat{\hat{\mu}}_X) = \frac{Y}{n} ,$$

the UMV unbiased estimate of p/q.

Next examine the effect of using these two approaches to estimate a different function of p, namely, the variance of X. The first is to use $\hat{p}$

to yield

$$\hat{\sigma}^2 = \hat{p}/\hat{q}^2 = \frac{Y(n + Y)}{n^2} .$$

The bias of $\hat{\sigma}^2$ can easily be shown to be

$$B(n,\sigma^2) = \left(\frac{1}{n}\right)\left(\frac{p}{q^2}\right)$$

and consequently the ordinary jackknife will produce an unbiased estimator. First we note that the second estimator for σ^2 in the jackknife method is given by

$$\hat{\theta}_2 = \frac{1}{n}\sum_{i=1}^{n} \frac{(Y - X_i)(Y - X_i + n - 1)}{(n - 1)^2}$$

$$= \frac{1}{n(n - 1)^2}\sum_{i=1}^{n} [Y^2 - 2X_iY + (n - 1)Y$$

$$+ X_i^2 - (n - 1)X_i]$$

$$= \frac{n - 2}{n(n - 1)^2} Y^2 + \frac{1}{n} Y + \frac{1}{n(n - 1)^2} \sum X_i^2 .$$

Consequently

$$J(\hat{\sigma}^2) = n\hat{\sigma}^2 - (n - 1)\hat{\theta}_2$$

$$= Y + \frac{Y^2}{n} - \frac{n - 2}{n(n - 1)} Y^2 - \frac{n - 1}{n} Y - \frac{1}{n(n - 1)} \sum X_i^2$$

$$= \frac{1}{n(n - 1)} Y^2 + \frac{1}{n} Y - \frac{1}{n(n - 1)} \sum X_i^2 \quad ,$$

or

$$J(\hat{\sigma}^2) = \frac{1}{n(n - 1)} \sum_{i \neq j}\sum X_i(X_j + 1) \quad .$$

If, however, we had chosen to use the corresponding function of $\hat{\hat{p}}$ to estimate the variance, then the estimator under consideration would be

$$\hat{\hat{\sigma}}^2 = \hat{\hat{p}}/\hat{\hat{q}}^2 = \frac{Y(n + Y - 1)}{(n - 1)^2} \quad .$$

It may be shown that

$$E[\hat{\hat{\sigma}}^2] = \frac{n}{n - 1} \frac{p}{q} + \frac{np + n^2p^2}{(n - 1)^2q^2}$$

$$= \frac{p}{q^2} + \frac{n}{(n - 1)^2}\left(\frac{2p + p^2}{q^2}\right) - \frac{1}{(n - 1)^2} \frac{p}{q^2} \quad ,$$

and here we have an example in which the proper $G^{(2)}$ estimator will completely eliminate the bias. In particular if we produce the estimators $\hat{\theta}_2$ and $\hat{\theta}_3$ in the standard way from $\hat{\theta}_1 = \hat{\hat{\sigma}}^2$, we have

$$G^{(2)}(\hat{\hat{\sigma}}^2) = \frac{\begin{vmatrix} \hat{\theta}_1 & \hat{\theta}_2 & \hat{\theta}_3 \\ \dfrac{n}{(n-1)^2} & \dfrac{n-1}{(n-2)^2} & \dfrac{n-2}{(n-3)^2} \\ -\dfrac{1}{(n-1)^2} & -\dfrac{1}{(n-2)^2} & -\dfrac{1}{(n-3)^2} \end{vmatrix}}{\begin{vmatrix} 1 & 1 & 1 \\ \dfrac{n}{(n-1)^2} & \dfrac{n-1}{(n-2)^2} & \dfrac{n-2}{(n-3)^2} \\ -\dfrac{1}{(n-1)^2} & -\dfrac{1}{(n-2)^2} & -\dfrac{1}{(n-3)^2} \end{vmatrix}} .$$

So, using the cofactor notation of the previous example

$$c_2^1 = \frac{(n-2) - (n-1)}{(n-2)^2(n-3)^2} = -\frac{1}{(n-2)^2(n-3)^2}$$

$$c_2^2 = \frac{n - n + 2}{(n - 1)^2(n - 3)^2} = \frac{2}{(n - 1)^2(n - 3)^2}$$

$$c_2^3 = \frac{n - 1 - n}{(n - 1)^2(n - 2)^2} = -\frac{1}{(n - 1)^2(n - 2)^2} ,$$

and

$$G^{(2)}(\hat{\hat{\sigma}}^2) = \frac{\sum_{k=1}^{3} c_2^k \, \hat{\theta}_k}{\sum_{k=1}^{3} c_2^k} ,$$

which simplifies to

$$G^{(2)}(\hat{\hat{\sigma}}^2) = \frac{1}{2}[(n - 1)^2\hat{\theta}_1 - 2(n - 2)^2\hat{\theta}_2 + (n - 3)^2\hat{\theta}_3] . \tag{4.10}$$

Familiarity with the coefficients of the second-order jackknife suggests that the bias of $(\hat{\hat{\sigma}}^2)$ may be reexpressed in a manner which (with practice) will allow one to form $G^{(2)}(\hat{\hat{\sigma}}^2)$ by inspection. In fact it is easy to see that we could have written

$$E[\hat{\hat{\sigma}}^2] = \frac{p}{q^2} + \frac{1}{(n - 1)} \frac{2p + p^2}{q^2} + \frac{1}{(n - 1)^2} \frac{p + p^2}{q^2} . \tag{4.11}$$

Moreover, simple algebra yields

$$J^{(2)}(\hat{\theta}) = \frac{1}{2}\,[\,n^2\hat{\theta}_1 - 2(n-1)^2\hat{\theta}_2 + (n-2)^2\hat{\theta}_3] \,. \tag{4.12}$$

Since $J^{(2)}(\hat{\theta})$ eliminates the first two terms of a power series expansion for the bias of $\hat{\theta}$, consideration of (4.11) and (4.12) would allow us to form (4.10) directly. In general it may be noted that to eliminate the first two terms of a power series expansion in $1/(n - s)$ the generalized jackknife takes the form

$$G^{(2)}(\hat{\theta}) = \frac{1}{2}\,[(n-s)^2\hat{\theta}_1 - 2(n-s-1)^2\hat{\theta}_2 + (n-s-2)^2\hat{\theta}_3] \,.$$

Continuing with the present example, we see the second estimator derived from $\hat{\hat{\sigma}}^2$ is given by

$$\hat{\theta}_2 = \frac{1}{n}\sum_{i=1}^{n} \frac{(Y - X_i)(Y - X_i + n - 2)}{(n-2)^2}$$

$$= \frac{1}{n(n-2)}\,Y^2 + \frac{n-1}{n(n-2)}\,Y + \frac{1}{n(n-2)^2}\sum X_i^2 \,.$$

Using the identities

$$\sum_{i<j}\sum (X_i + X_j) = (n - 1)Y$$

and

$$\sum_{i<j}\sum (X_i + X_j)^2 = Y^2 + (n - 2) \sum X_i^2 ,$$

we further obtain

$$\hat{\theta}_3 = \frac{2}{n(n-1)(n-3)^2} \sum_{i<j}\sum (Y - X_i - X_j)(Y - X_i - X_j + n - 3)$$

$$= \frac{2}{n(n-1)(n-3)^2} \left[\frac{n(n-1)}{2} Y^2 - 2(n-1) Y^2 + \frac{n(n-1)(n-3)}{2} Y + (n-2) \sum X_i^2 + Y^2 - (n-1)(n-3)Y \right]$$

$$= \frac{n^2 - 5n + 6}{n(n-1)(n-3)^2} Y^2 + \frac{n-2}{n(n-3)} Y$$

$$+ \frac{2(n-2)}{n(n-1)(n-3)^2} \sum X_i^2 \quad .$$

Substitution in (4.10) then yields

$$G^{(2)}(\hat{\hat{\sigma}}^2) = \frac{1}{n(n-1)} Y^2 + \frac{1}{n} Y - \frac{1}{n(n-1)} \sum X_i^2 \quad ,$$

which is identical to the estimator, $J(\hat{\sigma}^2)$, obtained earlier in this example. Some notice should be taken of the fact that in these closely related problems, a first-order and a second-order jackknife have produced the very same estimator. Thus we see that the higher order jackknife will not necessarily induce excessive variability.

Another interesting comparison to be made is with the UMV unbiased estimator $\tilde{\sigma}^2$ which can be found by modifying $\hat{\sigma}^2$. Note that we may write

$$\tilde{\sigma}^2 = \frac{Y(n+Y)}{n(n+1)}$$

$$= \frac{1}{n(n+1)} \sum_i \sum_j X_i(X_j + 1) \quad ,$$

$$J(\hat{\sigma}^2) = G^{(2)}(\hat{\hat{\sigma}}^2) = \frac{1}{n(n-1)} \sum_{i \neq j} \sum X_i(X_j + 1) \quad .$$

Therefore, although $J(\hat{\sigma}^2)$ and $G^{(2)}(\hat{\hat{\sigma}}^2)$ are not UMV, they are not radically different.

<u>Example 4.4.</u> Suppose that in the absence of any information about the bias we assume

$$f_{ij}(n) = (n - j + 1)^{-i} ,$$

and we wish to eliminate the first two terms of (4.6). This dictates the use of

$$J^{(2)}(\hat{\theta}) = \frac{1}{2} \left[n^2\hat{\theta}_1 - 2(n - 1)^2\hat{\theta}_2 + (n - 2)^2\hat{\theta}_3 \right] .$$

Let us examine the bias of $J^{(2)}(\hat{\theta})$ if the actual bias of the original estimator were given by

$$B(n,\theta) = \frac{a}{n^3} + \frac{b}{n^4} .$$

In this event

$$E[\hat{\theta}_2 - \theta] = \frac{a}{(n - 1)^3} + \frac{b}{(n - 1)^4} ,$$

$$E[\hat{\theta}_3 - \theta] = \frac{a}{(n - 2)^3} + \frac{b}{(n - 2)^4} .$$

Hence

$$\text{bias}\,[J^{(2)}(\hat{\theta})] = \frac{a}{2}\left[\frac{1}{n} - \frac{2}{n-1} + \frac{1}{n-2}\right]$$

$$+ \frac{b}{2}\left[\frac{1}{n^2} - \frac{2}{(n-1)^2} + \frac{1}{(n-2)^2}\right]$$

$$= \frac{a}{n(n-1)(n-2)} + \frac{6b}{n^2(n-2)^2}$$

$$- \frac{2b}{n^2(n-1)^2(n-2)^2} ,$$

which, for $n > 2$, represents a strict increase over the original bias. This hypothetical bias structure is not merely a pathological example but would arise from a previous application of the second-order ordinary jackknife to an estimator in which the bias is a power series expansion in $(1/n)$. This is a matter of some concern for those reapplication schemes (not discussed here) which disregard the known effect of the first application, i.e., we should be skeptical of $J^{(2)}[J^{(2)}(\hat{\theta})]$. Similar comments can be made concerning $J[J(\hat{\theta})]$.

In the above example we have seen as in the case of $G(\hat{\theta})$ that an improper choice of a_{ij} can actually result in an increase in bias. Since, again as in the case of $G(\hat{\theta})$, it is common practice to take $a_{ij} = [1/(n - j + 1)^i]$, when the $f_{ij}(n)$ are not known, we are again faced with the problem of characterizing the bias reduction properties of the estimator. We have of course already accomplished this to some extent when k = 1, by Theorem 3.3. The following theorems, which were first shown in [2], extend that result to k = 2 and similar extensions can be made by the reader.

THEOREM 4.3. Let

$$B(n,\theta) = E[\hat{\theta} - \theta] \quad .$$

Then

$$E[J^{(2)}(\hat{\theta})] = \theta + B(n,\theta) + (2n - 3)\ \Delta B(n,\theta)$$

$$+ \frac{(n - 2)^2}{2}\ \Delta^2 B(n,\theta) \quad , \qquad (4.13)$$

where

$$\Delta B(n,\theta) = B(n,\theta) - B(n - 1,\theta)$$

and

$$\Delta^2 B(n,\theta) = B(n,\theta) - 2B(n - 1,\theta) + B(n - 2,\theta) \quad .$$

Proof: Taking the expected value of $J^{(2)}(\hat{\theta})$ we have

$$\frac{\begin{vmatrix} \theta + B(n,\theta) & \theta + B(n-1,\theta) & \theta + B(n-2,\theta) \\ \dfrac{1}{n} & \dfrac{1}{n-1} & \dfrac{1}{n-2} \\ \dfrac{1}{n^2} & \dfrac{1}{(n-1)^2} & \dfrac{1}{(n-2)^2} \end{vmatrix}}{\begin{vmatrix} 1 & 1 & 1 \\ \dfrac{1}{n} & \dfrac{1}{n-1} & \dfrac{1}{n-2} \\ \dfrac{1}{n^2} & \dfrac{1}{(n-1)^2} & \dfrac{1}{(n-2)^2} \end{vmatrix}}$$

$$= \theta + B(n-1,\theta) + 2(n-1)[B(n-1,\theta) - B(n-2,\theta)]$$

$$+ \frac{n^2}{2}[B(n,\theta) - 2B(n-1,\theta) + B(n-2,\theta)]$$

$$= \theta + B(n,\theta) - [B(n,\theta) - B(n-1,\theta)]$$

$$+ 2(n-1)[B(n,\theta) - B(n-1,\theta) - B(n,\theta)$$

$$+ 2B(n-1,\theta) - B(n-2,\theta)] + \frac{n^2}{2}\Delta^2 B(n,\theta)$$

$$= \theta + B(n,\theta) + (2n-3)\Delta B(n,\theta) + \frac{(n-2)^2}{2}\Delta^2 B(n,\theta) \quad ,$$

as was to be shown.

We are now in a position to characterize the bias in $J^{(2)}(\hat{\theta})$ in the manner in which we previously characterized $J^{(1)}(\hat{\theta})$.

THEOREM 4.4. Let $B(n,\theta)$ be defined as in Theorem 4.3 and assume there exists a $p > 0$ such that

$$\lim_{n\to\infty} n^p B(n,\theta) = C(0) \neq 0 \quad \text{or } \pm \infty, \tag{4.14}$$

and $\lim_{n\to\infty} n^{p+2} \Delta^2 B(n,\theta)$ exists. Then

(i) if $p = 1$ or $p = 2$, then $J^{(2)}(\hat{\theta})$ L.O.B.E. $\hat{\theta}$,

(ii) if $p < 3$, $p \neq 1$, $p \neq 2$, then $J^{(2)}(\hat{\theta})$ B.S.O.B.E. $\hat{\theta}$,

(iii) if $p = 3$, then $J^{(2)}(\hat{\theta})$ S.O.B.E. $\hat{\theta}$,

(iv) if $p > 3$, then $\hat{\theta}$ B.S.O.B.E. $J^{(2)}(\hat{\theta})$.

Proof: As before

$$\lim_{n\to\infty} \frac{B(n,\theta)}{n^{-p}} = \lim_{n\to\infty} \frac{\Delta B(n,\theta)}{\Delta n^{-p}} = C(\theta)$$

and hence $\Delta B(n,\theta) \to 0$ as $n \to \infty$. Moreover the existence of $\lim_{n\to\infty} n^{p+2} \Delta^2 B(n,\theta)$ implies the existence of $\lim_{n\to\infty} [\Delta^2 B(n,\theta)/\Delta^2(n^{-p})]$. Thus a second application of Theorem 3.1 gives

$$\lim_{n\to\infty} \frac{\Delta(\Delta B(n,\theta))}{\Delta(\Delta n^{-p})} = \lim_{n\to\infty} \frac{\Delta^2 B(n,\theta)}{\Delta^2(n^{-p})} = C(\theta) .$$

Now from Theorem 4.3 we have

$$E[J^{(2)}(\hat{\theta}) - \theta] = B(n,\theta) + (2n - 3)\Delta B(n,\theta) + \frac{(n - 2)^2}{2} \Delta^2 B(n,\theta) ,$$

and therefore

$$\left| \lim_{n\to\infty} \frac{E[J^{(2)}(\hat{\theta}) - \theta]}{E[\hat{\theta} - \theta]} \right|$$

$$= \left| \lim_{n\to\infty} \frac{B(n,\theta) + (2n - 3)\Delta B(n,\theta) + \frac{(n - 2)^2}{2} \Delta^2 B(n,\theta)}{B(n,\theta)} \right|$$

$$= \left| 1 + \lim_{n\to\infty} \frac{(2n - 3)\Delta B(n,\theta)}{B(n,\theta)} + \lim_{n\to\infty} \frac{(n - 2)^2}{2} \frac{\Delta^2 B(n,\theta)}{B(n,\theta)} \right| . \tag{4.15}$$

Now by the proof of Theorem 3.3 we have

$$\lim_{n\to\infty} \frac{(2n - 3)\Delta B(n,\theta)}{B(n,\theta)} = -2p \quad .$$

Moreover

$$\lim_{n\to\infty} \frac{(n - 2)^2}{2} \frac{\Delta^2 B(n,\theta)}{B(n,\theta)}$$

$$= \lim_{n\to\infty} \left\{ \frac{(n - 2)^{p+2}\Delta^2(n^{-p})[\Delta^2 B(n,\theta)/\Delta^2(n^{-p})]}{2(n - 2)^p B(n,\theta)} \right\}$$

$$= \lim_{n\to\infty} \left\{ \frac{(n - 2)^{p+2}\Delta^2(n^{-p})C(\theta)}{2C(\theta)} \right\}$$

$$= \frac{p(p + 1)}{2} \quad .$$

Consequently by (4.15)

$$\left| \lim_{n\to\infty} \frac{E[J^{(2)}(\hat{\theta}) - \theta]}{E[\hat{\theta} - \theta]} \right| = \left| 1 - 2p + \frac{p(p + 1)}{2} \right|$$

$$= \frac{|(p - 1)(p - 2)|}{2} \quad . \qquad (4.16)$$

The result follows immediately. Example 4.1 provides an illustration of case (i) of this theorem, while Example 4.4 demonstrates case (iii).

In the above theorems we have shown that in many cases $J^{(2)}$ will be an effective bias removal tool. Moreover it should now be clear that $J^{(2)}$ can be more effective than $J^{(1)}$. To establish an asymptotic characterization of this property we have the next two theorems.

THEOREM 4.5. Let $B(n,\theta)$ and p be defined as in Theorem 4.4 and assume $p \neq 1$. Then

(i) if $p = 2$, $J^{(2)}(\hat{\theta})$ L.O.B.E. $J(\hat{\theta})$,

(ii) if $p < 4$, $p \neq 2$, $J^{(2)}(\hat{\theta})$ B.S.O.B.E. $J(\hat{\theta})$,

(iii) if $p = 4$, $J^{(2)}(\hat{\theta})$ S.O.B.E. $J(\hat{\theta})$,

(iv) if $p > 4$, $J(\hat{\theta})$ B.S.O.B.E. $J^{(2)}(\hat{\theta})$.

Proof:

$$\frac{E[J^{(2)}(\hat{\theta}) - \theta]}{E[J(\hat{\theta}) - \theta]}$$

$$= \frac{B(n,\theta) + (2n - 3)\Delta B(n,\theta) + [(n - 2)^2/2]\Delta^2 B(n,\theta)}{B(n,\theta) + (n - 1)\Delta B(n,\theta)} \tag{4.17}$$

$$= \frac{[B(n,\theta) + (2n - 3)\Delta B(n,\theta) + [(n - 2)^2/2]\Delta^2 B(n,\theta)]/B(n,\theta)}{[B(n,\theta) + (n - 1)\Delta B(n,\theta)]/B(n,\theta)}$$

Now taking the limit of both sides of (4.17) we have

$$\left|\lim_{n\to\infty} \frac{E[J^{(2)}(\hat{\theta}) - \theta]}{E[J(\hat{\theta}) - \theta]}\right| = \frac{1}{2}\left|\frac{(p - 1)(p - 2)}{1 - p}\right|$$

$$= \frac{1}{2}\left| p - 2\right|. \tag{4.18}$$

The result then follows by examining equation (4.18).

In the above theorem it was necessary to assume $p \neq 1$. This particular restriction is eliminated by the next theorem.

<u>THEOREM 4.6</u>. Let $B(n,\theta)$ and p be defined as in Theorem 4.4 and assume $p = 1$. If there exists a $k_1 > 0$ such that

$$\lim_{n\to\infty} n^{k_1} [nB(n,\theta) - C(\theta)] = C_1(\theta) \neq 0 \quad \text{or } \pm\infty$$

and

$$\lim_{n\to\infty} n^{k_1+2} \Delta^2[nB(n,\theta)]$$

exists, then

(i) if $k_1 = 1$, $J^{(2)}(\hat{\theta})$ L.O.B.E. $J(\hat{\theta})$,

(ii) if $0 < k_1 < 3$ but $k_1 \neq 1$, $J^{(2)}(\hat{\theta})$ B.S.O.B.E. $J(\hat{\theta})$,

(iii) if $k_1 = 3$, $J^{(2)}(\hat{\theta})$ S.O.B.E. $J(\hat{\theta})$,

(iv) if $k_1 > 3$, $J(\hat{\theta})$ B.S.O.B.E. $J^{(2)}(\hat{\theta})$.

Proof: From our initial assumptions we have

$$C_1(\theta) = \lim_{n\to\infty} \frac{nB(n,\theta) - C(\theta)}{n^{-k_1}} = \lim_{n\to\infty} \frac{\Delta[nB(n,\theta) - C(\theta)]}{\Delta n^{-k_1}}$$

$$= \lim_{n\to\infty} \frac{nB(n,\theta) - (n-1)B(n-1,\theta)}{\Delta n^{-k_1}}$$

$$= \lim_{n\to\infty} \frac{B(n,\theta) + (n-1)\Delta B(n,\theta)}{\Delta n^{-k_1}} . \tag{4.19}$$

Furthermore since $p = 1$ the numerator in (4.19) goes to zero and therefore

$$C_1(\theta) = \lim_{n\to\infty} \frac{\Delta^2[nB(n,\theta) - C(\theta)]}{\Delta^2 n^{-k_1}}$$

$$= \lim_{n\to\infty} \frac{nB(n,\theta) - 2(n-1)B(n-1,\theta) + (n-2)B(n-2,\theta)}{\Delta^2 n^{-k_1}}$$

$$= \lim_{n\to\infty} \frac{2\Delta B(n,\theta) + (n-2)\Delta^2 B(n,\theta)}{\Delta^2 n^{-k_1}} .$$

Now consider

$$\left| \lim_{n\to\infty} \frac{B(n,\theta) + (2n-3)\Delta B(n,\theta) + [(n-2)^2/2]\Delta^2 B(n,\theta)}{B(n,\theta) + (n-1)\Delta B(n,\theta)} \right|$$

$$= \frac{1}{2} \left| 2 + \lim_{n\to\infty} \frac{2(n-2)\Delta B(n,\theta) + (n-2)^2 \Delta^2 B(n,\theta)}{B(n,\theta) + (n-1)\Delta B(n,\theta)} \right|$$

$$= \frac{1}{2} \left| 2 + \lim_{n\to\infty} \frac{(n-2)(\Delta^2 n^{-k_1}) \left[\dfrac{2\Delta B(n,\theta) + (n-2)\Delta^2 B(n,\theta)}{\Delta^2 n^{-k_1}} \right]}{\Delta n^{-k_1} \left[\dfrac{B(n,\theta) + (n-1)\Delta B(n,\theta)}{\Delta n^{-k_1}} \right]} \right|$$

$$= = \frac{1}{2} \left| 2 + (-k_1 - 1) \frac{C_1(\theta)}{C_1(\theta)} \right|$$

$$= \frac{1}{2} \left| k_1 - 1 \right| , \qquad (4.20)$$

since

$$\lim_{n\to\infty} \frac{(n-2)\Delta^2 (n^{-k_1})}{\Delta n^{-k_1}} = -(k_1 + 1) .$$

The theorem now easily follows by examination of (4.20).

In concluding this section we should mention one more result which undoubtedly has already crossed the reader's mind. Namely, if it so happens that $J^{(1)}(\hat{\theta})$ is unbiased (even though this may be unknown to the statistician) will one induce bias by using $J^{(2)}(\hat{\theta})$? The answer to the question is in the negative, i.e., if $J^{(1)}(\hat{\theta})$ is unbiased for θ, then $J^{(2)}(\hat{\theta})$ is also unbiased for θ. The proof of this result is not difficult and much like the proofs above. It can be found in [2]. These brief comments suggest the next section.

5. Estimators Invariant Under Jackknifing

In the previous pages we have discussed the manner in which the jackknife reduces the bias of an estimator. In this final section we would like to show that for certain families of distributions the jackknife completely characterizes unbiased estimators. That is, we will show that under certain conditions an estimator is unbiased if and only if it is invariant under the jackknife, i.e., if and only if $J(\hat{\theta}) \equiv \hat{\theta}$. This is the content of the next theorem. In this theorem we use the symbol $J^{(0)}(\hat{\theta})$ to denote $\hat{\theta}$.

<u>THEOREM 5.1.</u> Let $\hat{\theta}$ be an estimator for θ defined on a random sample of size n and let

$$E[\hat{\theta}] = \theta + B(n,\theta) \tag{5.1}$$

where $B(n,\theta)$ is a function of n and θ such that

$$\lim_{n\to\infty} B(n,\theta) = 0.$$

Now if $k = 0$ or 1, $J^{(k+1)}(\hat{\theta})$ is defined for $n > n_0$, and

$$J^{(k+1)}(\hat{\theta}) = J^{(k)}(\hat{\theta}) \quad \text{for all } n > n_0 , \tag{5.2}$$

then $J^{(k)}(\hat{\theta})$ is an unbiased estimator for θ. Moreover, if for $n > n_0$

$$J^{(k+1)}(\hat{\theta}) \not\equiv J^{(k)}(\hat{\theta}) \quad , \tag{5.3}$$

and $J^{(k+1)}(\hat{\theta}) - J^{(k)}(\hat{\theta})$ is complete, then $J^{(k)}(\hat{\theta})$ is a biased estimator for θ.

Proof:

Case 1: $k = 0$. Suppose (5.2) holds. Then by Theorem 3.2

$$B(n,\theta) = B(n,\theta) + (n - 1)\Delta B(n,\theta),\ n > n_0 .$$

Therefore

$$\Delta B(n,\theta) \equiv 0 \quad ,$$

which implies

$$B(n,\theta) = C_2(\theta) = 0 \quad ,$$

and hence $\hat{\theta}$ is unbiased.

Now suppose (5.3) holds and $\hat{\theta}$ is unbiased. Since $\hat{\theta}$ is unbiased we know $J(\hat{\theta})$ is unbiased and hence

$$E[J(\hat{\theta}) - \hat{\theta}] = 0$$

for all n and θ. But since $J(\hat{\theta}) - \hat{\theta}$ is complete we have

$$J(\hat{\theta}) \equiv \hat{\theta} \quad .$$

This is a contradiction and hence $\hat{\theta}$ is biased.

Case 2: k = 1. Again suppose (5.2) holds. Then

$$B(n,\theta) + (n - 1)\Delta B(n,\theta)$$

$$= B(n,\theta) + (2n - 3)\Delta B(n,\theta) + [(n - 2)^2/2]\Delta^2 B(n,\theta),$$

for $n > n_0$. Consequently,

$$2\Delta B(n,\theta) + (n - 1)\Delta^2 B(n,\theta) = 0 \quad ,$$

and therefore

$$B(n,\theta) = \frac{C_1(\theta)}{n} + C_2(\theta)$$

$$= \frac{C_1(\theta)}{n} \quad ,$$

since $C_2(\theta)$ must be zero. Thus $J(\hat{\theta})$ is unbiased. The remaining part of the proof follows in exactly the same manner as Case I and thus the theorem is established. The condition $\lim_{n\to\infty} B(n,\theta) = 0$ is not necessary in the above theorem but it has been included because it makes the theorem easier to state correctly.

Examples 3.1 and 3.2 provide illustrations of this theorem. Recall that when

$$\hat{\theta}_1 = \frac{1}{n} \sum_{i=1}^{n} (X_i - \overline{X})^2 ,$$

it followed that

$$\hat{\theta}_2 = \frac{n(n - 2)}{(n - 1)^2} \hat{\theta}_1 ,$$

and consequently

$$J(\hat{\theta}_1) = \frac{n}{n - 1} \hat{\theta}_1 .$$

Hence it is obvious that

$$J^{(1)}(\hat{\theta}_1) \neq J^{(0)}(\hat{\theta}_1) = \hat{\theta}_1$$

and, at least when the X's are normal, the distribution of the difference

$$J^{(1)}(\hat{\theta}_1) - \hat{\theta}_1 = \frac{1}{n-1} \hat{\theta}_1$$

is complete; therefore we are assured that $\hat{\theta}_1$ is biased.

Next, it can be shown that

$$\hat{\theta}_3 = \frac{2}{n(n-1)} \sum_{i<j}\sum \left[\frac{1}{n-2} \sum_{k\neq i,j} X_k^2 - \left(\frac{n\bar{X} - X_i - X_j}{n-2} \right)^2 \right]$$

$$= \frac{n(n-3)}{(n-1)(n-2)} \hat{\theta}_1 \;.$$

Hence

$$J^{(2)}(\hat{\theta}_1) = \frac{1}{2} [n^2\hat{\theta}_1 - 2(n-1)^2\hat{\theta}_2 + (n-2)^2\hat{\theta}_3]$$

$$= \frac{1}{2} [n^2 - 2n(n-2) + \frac{n(n-2)(n-3)}{n-1}]\hat{\theta}_1$$

$$= \frac{n}{n - 1} \hat{\theta}_1$$

$$= J(\hat{\theta}_1)$$

and by Theorem 5.1 we may conclude that $J(\hat{\theta}_1)$ is unbiased.

In Example 3.2 the estimator

$$\hat{p}^2 = \left(\frac{X}{n}\right)^2$$

was considered and the result was

$$J(\hat{p}^2) = \frac{X(X - 1)}{n(n - 1)} \quad .$$

In a similar fashion we note that if two distinct trials are deleted from the sample, the estimator assumes the values

$$\left(\frac{x - 2}{n - 2}\right)^2, \quad \left(\frac{x - 1}{n - 2}\right)^2, \quad \text{and} \quad \left(\frac{x}{n - 2}\right)^2$$

and does so with frequencies

$$\frac{\binom{x}{2}}{\binom{n}{2}}, \quad \frac{x(n-x)}{\binom{n}{2}}, \quad \text{and} \quad \frac{\binom{n-x}{2}}{\binom{n}{2}},$$

respectively. Hence, the third estimator is

$$\frac{2X}{(n-2)^2 n(n-1)} \left[\frac{(X-1)(X-2)^2}{2} + (n-X)(X-1)^2 + \frac{X(n-X)(n-X-1)}{2} \right]$$

$$= \frac{(n-3)X^2 + 2X}{n(n-1)(n-2)} .$$

Therefore the second-order jackknife is given by

$$J^{(2)}(\hat{p}^2) = \frac{1}{2}\left[n^2 \frac{X^2}{n^2} - 2(n-1)^2 \frac{(n-2)X^2 + X}{n(n-1)^2} + (n-2)^2 \frac{(n-3)X^2 + 2X}{n(n-1)(n-2)} \right]$$

$$= \frac{X(X - 1)}{n(n - 1)}$$

$$= J(\hat{p}^2) \quad ,$$

and consequently the first-order jackknife is unbiased.

From an alternative point of view suppose that we are considering the estimator

$$\tilde{p}^2 = \frac{X(X - 1)}{n(n - 1)} \quad .$$

The second estimator, which Quenouille's procedure yields, when all subsets of size (n - 1) are averaged, is

$$\frac{1}{n}\left[X\frac{(X - 1)(X - 2)}{(n - 1)(n - 2)} + (n - X)\frac{X(X - 1)}{(n - 1)(n - 2)}\right]$$

$$= \frac{X(X - 1)}{n(n - 1)} \quad .$$

It is then trivially true that

$$J(\tilde{p}^2) = \frac{X(X - 1)}{n(n - 1)}$$

and this again yields the conclusion that the estimator $X(X - 1)/n(n - 1)$ is unbiased.

CHAPTER II
APPLICATIONS TO BIASED ESTIMATORS

1. Introduction

Most of the example applications of the generalized jackknife given in Chapter I were purely illustrative and consequently many were of a trivial nature. In this chapter we shall examine some nontrivial applications where additional information is incorporated in the G estimator through the parameter R. We first consider the application of the jackknife and the generalized jackknife to problems in ratio estimation. Following this presentation the general problem of estimating a truncation point is analyzed and finally the manuscript problem of Mosteller and Tukey [32] is considered as an example from discriminant analysis.

2. Ratio Estimators

In the theory of sampling there is a strong emphasis placed upon the use of auxiliary information. An example of this fact which is of interest here is the use of auxiliary information to improve the precision of estimates through consideration of a population ratio, say ρ. More specifically, often a situation exists where the ratio of a variable Y to another variable X is believed to have a smaller variance than the Y variable alone. Suppose, for example, one were interested in the value of a population total T(Y). Rather than estimate this total directly from the sample it may be better to estimate $\rho = T(Y)/T(X)$ from the sample and then multiply it by the known total, T(X), to estimate the total T(Y). This is called ratio estimation. An instructive hypothetical example is given by Raj [35] as follows.

Suppose it is desired to estimate the total agricultural area, T(Y), in a region containing N communes. There are very big communes and very small communes which causes Y, the agricultural area of an individual commune, to vary tremendously over the region. But the ratio, Y/X of agricultural area to the population of the commune (per capita area) would be less variable. If the population figures, X, are known for each commune, it would be preferable to estimate the ratio of agricultural area to the census population from the sample of communes and then multiply this figure by the known census population total, T(X), of all the communes in the region. Thus, if a random sample of n communes gives

$\sum_{i=1}^{n} Y_i$ and $\sum_{i=1}^{n} X_i$ as the sample totals for area

and population respectively, the estimator for T(Y) is

$$\hat{T}(Y) = T(X) \sum_{i=1}^{n} Y_i \Big/ \sum_{i=1}^{n} X_i \quad , \qquad (2.1)$$

where T(X) is the known total population for the region. If the census data on X is not employed, then one might use the estimator

$$\hat{\hat{T}}(Y) = N \sum_{i=1}^{n} Y_i \Big/ n \quad ,$$

which should have a larger variance than $\hat{T}(Y)$ in our hypothetical case.

The ratio estimator (2.1) is biased, though in many situations only negligibly so. On the other hand the bias may be considerable in surveys with many strata or small or moderate samples within strata if it is deemed appropriate to use separate ratio estimators for each stratum. When it is considered to be important that proper confidence statements be made, it is often necessary that the bias of an estimator be negligibly small. Consequently, in recent years, considerable attention has been given to the development of unbiased or approximately unbiased ratio estimators.

The following theorem shows that

$$r = \sum_{i=1}^{n} Y_i \Big/ \sum_{i=1}^{n} X_i = \bar{Y} \Big/ \bar{X}$$

is usually a biased estimator of ρ.

<u>THEOREM 2.1</u>. In simple random sampling, the bias of the ratio estimator r is given by

$$E[r] - \rho = -[E(\bar{X})]^{-1} \mathrm{cov}\ (r,\bar{X}) \quad .$$

Note that the bias associated with the estimator for the total of Y,

$$\hat{T}(Y) = T(X)(\overline{Y}/\overline{X}) \quad ,$$

is given by

$$\text{bias } [\hat{T}(Y)] = T(X) \text{ bias } (r).$$

Also note that practically speaking r is never unbiased since this occurs only if r and $\overline{X}$ are uncorrelated, a situation which seldom occurs in practice.

The decision to use a ratio estimator in hopes of improving the precision is ordinarily based on consideration of the coefficients of variation for the variables X and Y and upon the correlation believed to be present between the two. In general the ratio estimator is useful if the characters X and Y have a correlation coefficient which exceeds 1/2. The variable Y must be nearly proportional to X or other schemes such as difference estimators or regression estimators may be dictated.

After the decision to use a ratio estimator has been made, the evaluations of the various modifications to the classical estimator which exist will, of necessity, depend upon the assumed model for the relationship between Y and X and for the family to which the distribution of X belongs. Several authors have examined the bias and approximations to the mean square error (MSE) of various estimators under two general models.

Durbin [17] examined ratio estimators of the form $r = \overline{Y}/\overline{X}$, where the regression of Y on X is linear and X is normally distributed. He considers an application of Quenouille's method which splits the sample into two equal size sets to yield

$$r_1 = \overline{Y}_1 / \overline{X}_1 \quad \text{and} \quad r_2 = \overline{Y}_2 / \overline{X}_2 \quad ,$$

where

$$\overline{Y} = \frac{1}{2}(\overline{Y}_1 + \overline{Y}_2)$$

$$\overline{X} = \frac{1}{2}(\overline{X}_1 + \overline{X}_2) \quad .$$

Then the new estimate, $\hat{\rho}_2$, of $\rho = E(Y)/E(X)$ is

$$\hat{\rho}_2 = 2r - \frac{1}{2}(r_1 + r_2) \quad .$$

In addition Durbin investigated r and $\hat{\rho}_2$ assuming X has a gamma distribution. These two models form the basis for most of the work done in the area of bias reduction for ratio-type estimation.

Suppose

$$\overline{Y} = a + b\overline{X} + U \quad ,$$

where the var (U) = δ, a nonrandom quantity of $O(n^{-1})$, and $E[U|\overline{x}] = 0$. Hence,

$$\rho = E(Y)/E(X) = b + a/E(\bar{X}) \quad ,$$

and since $E[U|\bar{x}] = 0$,

$$E(r) = aE(\bar{X}^{-1}) + b \quad . \tag{2.2}$$

Consequently the bias in r is determined by the degree to which $E[\bar{X}^{-1}]$ differs from $(E[\bar{X}])^{-1}$.

2.1 Normal Auxiliary

Suppose that $\bar{X}$ is a normal variable with variance h, which is $0(n^{-1})$, and the units of measurement are chosen so that $E(\bar{X}) = 1$. Then let $\bar{X} = 1 - \xi$ and hence for sufficiently large n we have

$$E[\bar{X}^{-1}] = E(1 + \xi + \xi^2 + \xi^3 + \ldots) \quad .$$

Taking the first four nonvanishing terms we find

$$E[\bar{X}^{-1}] = 1 + h + 3h^2 + 15h^3 + 0(n^{-4}) \quad . \tag{2.3}$$

Similarly

$$E[\bar{X}^{-2}] = E(1 + 2\xi + 3\xi^2 + 4\xi^3 + \ldots)$$

$$= 1 + 3h + 15h^2 + 105h^3 + 0(n^{-4}) \quad . \tag{2.4}$$

At this point we see that if (2.3) is substituted in (2.2) the bias in r may be determined as

$$E(r) - \rho = aE(\bar{X}^{-1}) + b - (a + b)$$

$$= a(h + 3h^2 + 15h^3) \quad ,$$

neglecting terms of $0(n^{-4})$. Further, since $\mathrm{var}(\bar{X}_1) = \mathrm{var}(\bar{X}_2) = 2h$, we may replace h by 2h in (2.3) and (2.4) and obtain

$$E[\bar{X}_i^{-1}] = 1 + 2h + 12h^2 + 120h^3$$

and

$$E[\bar{X}_i^{-2}] = 1 + 6h + 60h^2 + 840h^3 \quad , \quad i = 1,2 \ .$$

Thus if $\hat{\theta}_2 = (r_1 + r_2)/2$, then its bias is given by

$$E(\hat{\theta}_2) - \rho = a(2h + 12h^2 + 120h^3) \quad .$$

Hence if the principles introduced in the previous chapter are employed to obtain a proper combination of

$$\hat{\theta}_1 = r$$

and

$$\hat{\theta}_2 = \frac{1}{2}(r_1 + r_2) \quad ,$$

of the form

$$\frac{\hat{\theta}_1 - R\hat{\theta}_2}{1 - R} \quad ,$$

then to eliminate the bias to terms of order $0(n^{-4})$ the appropriate choice of R is

$$R = \frac{1 + 3h + 15h^2}{2(1 + 6h + 60h^2)} \quad . \tag{2.5}$$

Selecting $R = 1/2$ leads to the estimator $\hat{\rho}_2$ studied by Durbin. For small h the above expression for R is quite near 1/2. The importance of (2.5) becomes clear when it is recalled that the auxiliary character X is being used because its distribution is known. Therefore h is not an unknown and this improved value for R may be used to yield

$$\hat{\rho}_3 = \frac{2(1 + 6h + 60h^2)r - (1 + 3h + 15h^2)(r_1 + r_2)/2}{1 + 9h + 105h^2}$$

Using (2.4) Durbin has shown that, not only is the bias of $\hat{\rho}_2$ smaller than that of r, but var $(\hat{\rho}_2)$ < var (r). The estimator $\hat{\rho}_3$ combines the same two quantities $\hat{\theta}_1$ and $\hat{\theta}_2$, in the same fashion with a further improvement in the bias. J. N. K. Rao [36] has shown that the bias and variance of the jackknifed classical ratio estimator are both decreasing functions of the number (N) of subsets into which the sample is split. In other words, let the sample of pairs (Y_i, X_i) $(i = 1,\ldots,n)$ be split at random into N groups each of size M. Then we get the estimator

$$\hat{\rho}^j = \overline{Y}^j / \overline{X}^j$$

from the sample after omitting the jth group, where

$$\overline{Y}^j = (n\overline{Y} - M\overline{Y}_j)/(n - M) ,$$

$$\overline{X}^j = (n\overline{X} - M\overline{X}_j)/(n - M) ,$$

and $\overline{Y}_j$ and $\overline{X}_j$ are the sample means for the jth group. Then Quenouille's estimator is

$$\hat{\rho}_Q = Nr - \frac{N - 1}{N} \sum_{1}^{N} \hat{\rho}^j , \tag{2.6}$$

and bias $(\hat{\rho}_Q)$ and var $(\hat{\rho}_Q)$ are both decreasing functions of N. For N = 2 $\hat{\rho}_Q$ becomes $\hat{\rho}_2$, the estimator given and studied by Durbin. Consequently, the indicated optimal choice of $\hat{\theta}_2$ is (corresponding to N = n)

$$\hat{\theta}_2 = \frac{1}{n} \sum_{j=1}^{n} \hat{\rho}^j .$$

However, the appropriate value of R is not $(n - 1)/n$ although in much of the current literature the assumption that $R = (n - 1)/n$ is inherent. Since the value of h is assumed to be known and h is $O(n^{-1})$, we shall consider the case in which $h = c/n$ for a known constant c. This requires us to choose

$$R = \frac{a[h + 3h^2 + 15h^3]}{a[c/(n - 1) + 3\,c^2/(n - 1)^2 + 15\,c^3/(n - 1)^3]}$$

$$= \left(\frac{n - 1}{n}\right)^3 \left[\frac{n^2 + 3cn + 15c^2}{n^2 + (3c - 2)n + (15c^2 - 3c + 1)}\right],$$

as the proper parameter in the estimator $G(\hat{\theta}_1, \hat{\theta}_2)$. This yields

$$G(\hat{\theta}_1, \hat{\theta}_2) = \hat{\rho}_4 = \frac{\hat{\theta}_1 - R\hat{\theta}_2}{1 - R} .$$

Thus $\hat{\rho}_4$ would appear to be the best of the estimators which we are considering here. Nevertheless, because one of the purposes of this section is one of illustrating a procedure and its potential usefulness, rather than discussing the optimal ratio estimator further, we shall avoid the complications of $\hat{\rho}_4$ and pursue the easier problem of comparing $\hat{\rho}_3$ and $\hat{\rho}_2$.

The following results, first given by Durbin, will be useful for this comparison:

$$\text{bias } (r) = a(h + 3h^2 + 15h^3) \triangleq aB(r)$$

$$\text{var } (r) = a^2(h + 8h^2 + 69h^3) + \delta(1 + 3h + 15h^2 + 105h^3)$$

$$\triangleq a^2 S_1(r) + \delta S_2(r)$$

$$\text{bias } (\hat{\rho}_2) = a(6h^2 + 90h^3) \triangleq aB(\hat{\rho}_2)$$

$$\text{var } (\hat{\rho}_2) = a^2(h + 4h^2 + 12h^3) + \delta(1 + 2h + 8h^2 + 108h^3)$$

$$\triangleq a^2 S_1(\hat{\rho}_2) + \delta S_2(\hat{\rho}_2) .$$

The estimator $\hat{\rho}_3$ is unbiased to $0(n^{-4})$. In order to derive the variance of $\hat{\rho}_3$ let

$$\hat{\rho}_3 = cr - d \frac{r_1 + r_2}{2} ,$$

where $c = 1/(1 - R)$, $d = R/(1 - R)$; $c - d = 1$. (Note that for $\hat{\rho}_2$ we have $R = 1/2$, $c = 2$, $d = 1$.)

Using the linear model introduced previously and splitting the sample as before we have $U = 1/2(U_1 + U_2)$, $\overline{Y}_i = a + b\overline{X}_i + U_i$, and $r_i = \overline{Y}_i/\overline{X}_i$, $i = 1,2$, and further that $E(U_i|\overline{x}_i) = 0$ and $E(U_i^2|\overline{x}_i) = 2\delta$.

Now we may write

$$\hat{\rho}_3 = cb + \frac{c}{\overline{X}}(a + U) - db - \frac{d}{2\overline{X}_1}(a + U_1) - \frac{d}{2\overline{X}_2}(a + U_2)$$

$$= b + a\left\{\frac{c}{\bar{X}} - \frac{d}{2}\left(\frac{1}{\bar{X}_1} + \frac{1}{\bar{X}_2}\right)\right\} + \frac{cU}{\bar{X}}$$

$$- \frac{d}{2}\left(\frac{U_1}{\bar{X}_1} + \frac{U_2}{\bar{X}_2}\right) .$$

Hence

$$E(\hat{\rho}_3 - b) = aE\left[\frac{c}{\bar{X}} - \frac{d}{2}\left(\frac{1}{\bar{X}_1} + \frac{1}{\bar{X}_2}\right)\right], \qquad (2.7)$$

and when R is given by (2.5)

$$c = \frac{2(1 + 6h + 60h^2)}{1 + 9h + 105h^2} ,$$

$$d = \frac{1 + 3h + 15h^2}{1 + 9h + 105h^2}$$

and therefore

$$E(\hat{\rho}_3 - b) = a + 0(n^{-4}) .$$

Next consider

$$E\left\{\frac{c}{\overline{X}} - \frac{d}{2}\left(\frac{1}{\overline{X}_1} + \frac{1}{\overline{X}_2}\right)\right\}^2$$

$$= E\left\{\frac{c^2}{\overline{X}^2} + \frac{d^2}{4}\left(\frac{1}{\overline{X}_1^2} + \frac{1}{\overline{X}_2^2}\right) + \left(\frac{d^2}{2} - cd\right)\left(\frac{1}{\overline{X}_1\overline{X}_2}\right)\right\}$$

$$= c^2(1 + 3h + 15h^2 + 105h^3)$$

$$+ \frac{d^2}{2}(1 + 6h + 60h^2 + 840h^3)$$

$$+ \left(\frac{d^2}{2} - 2cd\right)(1 + 2h + 12h^2 + 120h^3)^2 , \tag{2.8}$$

and

$$E_{\overline{X}}\, E_{U|\overline{X}}\left[\left\{\frac{c(U_1+U_2)}{2\overline{X}} - \frac{d}{2}\left(\frac{U_1}{\overline{X}_1} + \frac{U_2}{\overline{X}_2}\right)\right\}^2 \middle| \overline{X}_1, \overline{X}_2\right]$$

$$= E_{\overline{X}}\, 2\delta\left[\left(\frac{c}{2\overline{X}} - \frac{d}{2\overline{X}_1}\right)^2 + \left(\frac{c}{2\overline{X}} - \frac{d}{2\overline{X}_2}\right)^2\right]$$

$$= \frac{\delta}{2} E\left[\frac{2c^2}{\overline{X}^2} + d^2\left(\frac{1}{\overline{X}_1^2} + \frac{1}{\overline{X}_2^2}\right) - 4cd\left(\frac{1}{\overline{X}_1\overline{X}_2}\right)\right]$$

$$= \delta[c^2(1 + 3h + 15h^2 + 105h^3)$$

$$+ d^2(1 + 6h + 60h^2 + 840h^3)$$

$$- 2cd(1 + 2h + 12h^2 + 120h^3)]. \qquad (2.9)$$

Hence, substituting approximate expressions for c^2, d^2, and cd, we obtain, after some algebra,

$$E(\hat{\rho}_3 - b)^2 = a^2 \left[\frac{1 + 26h + 435h^2 + 4098h^3}{1 + 18h + 291h^2 + 1890h^3}\right]$$

$$+ \delta \left[\frac{1 + 28h + 471h^2 + 4680h^3}{1 + 18h + 291h^2 + 1890h^3}\right].$$

Consequently

$$\mathrm{var}\ (\hat{\rho}_3) = \mathrm{var}\ (\hat{\rho}_3 - b)$$

$$= a^2 \left[\frac{1 + 8h + 117h^2 + 2046h^3}{1 + 18h + 291h^2 + 1890h^3}\right]$$

$$+ \delta \left[\frac{1 + 28h + 471h^2 + 4680h^3}{1 + 18h + 291h^2 + 1890h^3}\right].$$

Since direct comparisons of the variance expressions are difficult the values of the coefficients, S_1 and S_2 of a^2 and δ, respectively, have been tabulated for several values of h. The coefficients of a in the expressions for the bias are also given here in Table 2.1. No bias was calculated for $\hat{\rho}_3$ since all terms containing the fourth power and higher in h have been neglected in the original approximations and $\hat{\rho}_3$ is corrected for bias to this degree. Because of the approximation, the entries in the lower half of the table are subject to considerable error. For instance the error in B(r) for h = 0.50 is greater than 6.0. In spite of this, the large values of h have been included for two reasons. First of all these larger values are not unusual in ordinary practical applications. Evidence in support of this contention may be found in J. N. K. Rao [37], where 16 natural populations are listed along with the coefficient of variation of the auxiliary variable X. Recall that h is the square of this coefficient. Six of the 16 populations exhibit a value of h greater than 1.0 and only two have a value of h less than 0.2. Second, note from Table 2.1 that even at h = 0.1 the bias in $\hat{\rho}_2$ is greater than that of r; a disturbing result since $\hat{\rho}_2$ was proposed from a bias reduction standpoint. However, recalling that the approximations made at the outset are valid only for small h, this latter observation is a commentary on the range of validity of these approximations rather than the jackknife method.

TABLE 2.1

VARIANCE COMPARISONS FOR NORMAL MODEL

h	r			$\hat{\rho}_2$			$\hat{\rho}_3$		
	B	S_1	S_2	B	S_1	S_2	B	S_1	S_2
0.01	0.010	0.011	1.032	0.001	0.010	1.021	0.0	0.077	1.100
0.05	0.058	0.079	1.201	0.016	0.061	1.133	0.0	0.331	1.454
0.10	0.145	0.249	1.555	0.150	0.152	1.388	0.0	0.528	1.736
0.15	0.268	0.563	2.142	0.439	0.280	1.844	0.0	0.646	1.900
0.20	0.440	1.072	3.040	0.960	0.456	2.584	0.0	0.722	2.005
0.25	0.672	1.828	4.328	1.781	0.687	3.687	0.0	0.776	2.078
0.30	0.975	2.883	6.085	2.97	0.984	5.236	0.0	0.815	2.130
0.40	1.84	6.096	11.320	6.72	1.808	9.992	0.0	0.870	2.203
0.50	3.125	11.125	19.375	12.75	3.000	17.500	0.0	0.906	2.250
0.75	8.765	34.359	55.984	41.34	8.062	52.562	0.0	0.958	2.318
1.00	19.0	78.0	124.0	96.0	17.0	119.0	0.0	0.987	2.355

Bias $(\cdot) = a \cdot B(\cdot)$. Variance $(\cdot) = a^2 S_1(\cdot) + \delta S_2(\cdot)$.

In each expression the terms of the fourth power in h and higher have been neglected.

The breakdown of the necessary approximation for the normal model for the realistically large coefficients of variation is one inducement to examine a different model. Furthermore, since practically all auxiliary random variables, X, which are used in real problems are positive, the normal model is not realistic when h is near 1. The gamma distribution has received more attention in the recent literature, possibly due to some of these arguments.

2.2 Gamma Auxiliary

Let (Y_i, X_i), $i = 1,\ldots,n$, denote a simple random sample from a population assumed to be infinite and let $\bar{Y}$ and $\bar{X}$ denote the sample means. Let $Y_i = a + bX_i + U_i$, where $E(U_i|x_i) = 0$ and $E(U_i^2|x_i) = n\delta$ and as before δ is a nonrandom quantity of $0(n^{-1})$. Let the variates X_i/n have the gamma distribution with parameter h so that $\bar{X}$ has the gamma distribution with parameter $m = nh$, i.e., the density of $\bar{X}$ is

$$\bar{x}^{\,m-1} \exp(-\bar{x})/\Gamma(m), \quad \bar{x} > 0.$$

It follows that

$$E\left[\frac{1}{\bar{X}}\right] = \frac{1}{m-1} \quad \text{and} \quad E\left[\frac{1}{\bar{X}^2}\right] = \frac{1}{(m-1)(m-2)} .$$

Proceeding as in the previous section we assume that n is even and the sample is randomly split in half.

Let $\overline{X}_j$ and $\overline{Y}_j$ $(j = 1,2)$ denote the sample means within the two groups. Hence $(1/2)\overline{X}_1$ and $(1/2)\overline{X}_2$ are independent gamma variables, each with parameter $m/2$, and

$$E\left[\frac{1}{\overline{X}_j}\right] = \frac{1}{m - 2} \quad \text{and} \quad E\left[\frac{1}{\overline{X}_j^2}\right] = \frac{1}{(m - 2)(m - 4)},$$

$j = 1,2$. Here we have $\rho = b + a/m$ and since

$$E\left[\frac{\overline{Y}}{\overline{X}}\right] - b = \frac{a}{m - 1},$$

the bias in r is

$$\text{bias}(r) = \frac{a}{m - 1} - \frac{a}{m} = \frac{a}{m(m - 1)}.$$

Furthermore the MSE of r is

$$\text{MSE}(r) = a^2\left[\frac{m + 2}{m^2(m - 1)(m - 2)}\right]$$

$$+ \delta\left[\frac{1}{(m - 1)(m - 2)}\right].$$

Now let

$$\hat{\theta}_2 = \frac{1}{2}(r_1 + r_2) = \frac{1}{2}\left(\frac{\bar{Y}_1}{\bar{X}_1} + \frac{\bar{Y}_2}{\bar{X}_2}\right)$$

as before, and then

$$E(\hat{\theta}_2 - b) = \frac{a}{2}\left(\frac{1}{m - 2} + \frac{1}{m - 2}\right) = \frac{a}{m - 2} .$$

Thus

$$\text{bias } (\hat{\theta}_2) = \frac{a}{m - 2} - \frac{a}{m} = \frac{2a}{m(m - 2)} .$$

Consequently, instead of choosing $R = 1/2$, which leads to $\hat{\rho}_2$ studied by Durbin and more recently by P. S. R. S. Rao [39], it is obvious that the proper choice is

$$R = \frac{\text{bias } (r)}{\text{bias } (\hat{\theta}_2)} = \frac{m - 2}{2(m - 1)} .$$

As in the previous section the result is

$$\hat{\rho}_3 = cr - d\left(\frac{r_1 + r_2}{2}\right) ,$$

where now $c = 2(m - 1)/m$ and $d = (m - 2)/m$.

It follows that $\hat{\rho}_3$ is unbiased for ρ and using the development of equations (2.8) and (2.9) we have

$$E(\hat{\rho}_3 - b)^2 = a^2 E\left\{ \frac{c^2}{\bar{X}^2} + \frac{d^2}{4}\left(\frac{1}{\bar{X}_1^2} + \frac{1}{\bar{X}_2^2}\right) + \left(\frac{d^2}{2} - 2cd\right)\left(\frac{1}{\bar{X}_1\bar{X}_2}\right)\right\}$$

$$+ \frac{\delta}{2} E\left\{ \frac{2c^2}{\bar{X}^2} + d^2\left(\frac{1}{\bar{X}_1^2} + \frac{1}{\bar{X}_2^2}\right) - 4cd\left(\frac{1}{\bar{X}_1\bar{X}_2}\right)\right\}$$

$$= a^2\left[\frac{c^2}{(m-1)(m-2)} + \frac{d^2}{2(m-2)(m-4)} + \frac{d^2/2 - 2cd}{(m-2)^2}\right] + \frac{\delta}{2}\left[\frac{2c^2}{(m-1)(m-2)} + \frac{2d^2}{(m-2)(m-4)} - \frac{4cd}{(m-2)^2}\right]$$

$$= \frac{a^2}{2m^2}\left[\frac{8(m-1)}{(m-2)} + \frac{(m-2)}{(m-4)} - \frac{7m^2 - 20m + 12}{(m-2)^2}\right] + \frac{\delta}{2m^2}\left[\frac{8(m-1)}{(m-2)} + \frac{2(m-2)}{m-4} - \frac{8m^2 - 24m + 16}{(m-2)^2}\right].$$

After more algebra we obtain

$$E(\hat{\rho}_3 - b)^2 = \frac{a^2(m-3)}{m^2(m-4)} + \frac{\delta(m-2)}{m^2(m-4)} ,$$

and therefore the MSE of $\hat{\rho}_3$ is

$$\text{MSE } (\hat{\rho}_3) = E(\hat{\rho}_3 - b)^2 - \frac{a^2}{m^2}$$

$$= \frac{a^2}{m^2(m-4)} + \frac{\delta(m-2)}{m^2(m-4)} .$$

Durbin compared the MSE of r to that of $\hat{\rho}_2$ and determined that

$$\text{MSE } (\hat{\rho}_2) < \text{MSE } (r)$$

provided that $m > 16$. Moreover the inequality might hold true for some values of m between 10 and 16. In order to attach some meaning to these values recall that

$$E(\bar{X}) = \text{var } (\bar{X}) = m$$

and hence the coefficient of variation of $\bar{X}$ is $m^{-1/2}$. Let us make this same comparison using the estimator

$\hat{\rho}_3$. Now

$$\text{MSE}(r) - \text{MSE}(\hat{\rho}_3) = \frac{a^2}{m^2}\left[\frac{(m+2)}{(m-1)(m-2)} - \frac{1}{(m-4)}\right] + \delta\left[\frac{1}{(m-1)(m-2)} - \frac{(m-2)}{m^2(m-4)}\right]$$

$$= \frac{a^2}{m^2}\left[\frac{(m-10)}{(m-1)(m-2)(m-4)}\right] + \frac{\delta}{m^2}\left[\frac{m^2 - 8m + 4}{(m-1)(m-2)(m-4)}\right],$$

which is positive for $m > 10$. Further, since the roots of $(m^2 - 8m + 4)$ are real, positive, and less than 8, the inequality

$$\text{MSE}(\hat{\rho}_3) < \text{MSE}(r)$$

may hold for some values of m between 8 and 10, depending upon the relative sizes of a and δ, and is surely valid for $m > 10$.

Further evidence of the superiority of $\hat{\rho}_3$ over

$\hat{\rho}_2$ is apparent in the comparison

$$\text{MSE }(\hat{\rho}_2) - \text{MSE }(\hat{\rho}_3) = \frac{a^2}{m^2}\left[\frac{m^3 - 5m^2 + 12m + 16}{(m - 1)(m - 2)^2(m - 4)} - \frac{1}{(m - 4)}\right] + \delta\left[\frac{m^2 - 7m + 18}{(m - 1)(m - 2)^2(m - 4)} - \frac{(m - 2)}{m^2(m - 4)}\right]$$

$$= \frac{a^2(4m + 20)}{m^2(m - 1)(m - 2)^2(m - 4)} + \frac{\delta(20m - 8)}{m^2(m - 1)(m - 2)^2(m - 4)}$$

which is positive over the entire range of reasonable values for m. Therefore, utilizing the generalized jackknife statistic we have realized a strict improvement in the bias and MSE of the ratio estimator under the assumed model.

2.3 Pseudoreplication

Another technique, peculiar to survey sampling, entitled pseudoreplication has been recently introduced and relates to the jackknife through the common usage of subsamples. The method, also referred to as half-sample replication, is a technique for the estimation of variance from stratified samples. We shall discuss it briefly here to point out the distinction between this method and the jackknife as applied to ratio estimation. This distinction will be helpful later when we discuss the use of the jackknife for variance estimation. For more on this topic the reader is referred to McCarthy [29].

If a set of primary sampling units is stratified to the extent that there are only two units drawn from each stratum, then there are only two independent replicates available for estimation of sampling precision. This results in interval estimates which are quite wide. Half-sample replication represents an attempt to alleviate this difficulty.

Consider a stratified sampling procedure where two independent observations are taken within each stratum. Suppose that the population and sample values are denoted by the following:

Stratum	Weight	Population mean	Population variance	Sample points	Sample mean
1	w_1	μ_1	σ_1^2	y_{11}, y_{12}	$\bar{y}_1$
2	w_2	μ_2	σ_2^2	y_{21}, y_{22}	$\bar{y}_2$
.	.	.	.	.	.
.	.	.	.	.	.
.	.	.	.	.	.
h	w_h	μ_h	σ_h^2	y_{h1}, y_{h2}	$\bar{y}_h$
.		.	.	.	.
.		.	.	.	.
.		.	.	.	.
L	w_L	μ_L	σ_L^2	y_{L1}, y_{L2}	$\bar{y}_L$

In the above $\sum_{h=1}^{L} w_h = 1$. An unbiased estimate of the population mean μ is given by

$$\bar{y} = \sum_{h=1}^{L} w_h \bar{y}_h ,$$

and the usual estimate of the variance of $\bar{Y}$, $V(\bar{Y})$, is

$$\hat{V}(\bar{y}) = \frac{1}{2} \sum_{h=1}^{L} w_h^2 s_h^2$$

$$= \frac{1}{4} \sum_{h=1}^{L} w_h^2 d_h^2$$

where $s_h^2 = \sum_{k=1}^{2} (y_{hk} - \bar{y}_h)^2$ and $d_h = (y_{h1} - y_{h2})$.

A half-sample replicate is defined as the sample obtained by selecting one of y_{11} and y_{12}, one of y_{21} and y_{22}, and so forth. There are 2^L distinct half-samples which can be drawn. The half-sample estimate of μ for the jth half-sample replicate is therefore

$$\bar{y}_j^* = \sum_{h=1}^{L} w_h \, y_{hi}$$

where i = 1 or 2 for each value of h. Clearly the average over all 2^L half-sample estimates is $\bar{y}$.

Now it can be shown that the deviations of the half-sample averages, $\bar{y}_j^*$, from the overall sample mean is of the form

$$(\bar{y}_j^* - \bar{y}) = \frac{1}{2}[\pm\, w_1 d_1 \pm w_2 d_2 \; \cdots \; \pm w_L d_L]$$

where the plus or minus sign for each term arises respectively as y_{h1} or y_{h2} is selected from stratum h for the particular half-sample under consideration. Hence

$$(\bar{y}_j^* - \bar{y})^2 = \frac{1}{4} \sum_{h=1}^{L} w_h^2 d_h^2 + \frac{1}{2} \sum_{h\,<}\sum_{k} \pm\, w_h w_k d_h d_k \quad ,$$

and because of the independence of selections within strata, $E[d_h d_k] = 0$, so that

$$E[(\overline{Y}^* - \overline{Y})^2] = \frac{1}{2} \sum_{h=1}^{L} w_h^2 \sigma_h^2 = V(\overline{Y}) .$$

These properties suggest the validity of the following approach to estimation of the variance of the sample mean. For fixed values of y_{hi} $(i = 1,2;\ h = 1,\ldots,L)$, the conditional population of all possible half-sample estimates, $\overline{y}_j^*$, has a population mean equal to $\overline{y}$. Furthermore the population average of $(\overline{y}_j^* - \overline{y})^2$ is $\hat{V}(\overline{y})$. Consequently, given a random sample of size k (sampling with replacement) from this population of 2^L half-sample means, we form

$$V^*(\overline{y}) = \frac{1}{k} \sum_{j=1}^{k} (\overline{y}_j^* - \overline{y})^2 ,$$

as an estimate of $V(\overline{Y})$. Properties of this estimator are discussed in [29].

The value of this technique lies in situations which are more complex than the simple linear one just presented. If one is faced with a more complicated sampling plan and, say, ratio-type estimators, then direct methods of analysis may either

(i) be unavailable,

(ii) require a prohibitive amount of computation relative to pseudoreplication, or

(iii) actually be inferior in some other facet to the method of half-samples.

It also happens that, if the pseudovalues of any jackknife are used to estimate the quantity $V(\overline{Y})$, the result is not different than the usual estimate $\hat{V}(\bar{y})$. Hence the jackknife method does not provide an alternative in the simple linear situations. However, in the more complex situations alluded to above, the jackknife method may, for reasons similar to (i) through (iii), provide an attractive alternative procedure for variance estimation. The subject of variance estimation and approximate confidence intervals through the use of pseudovalues of the jackknife will be taken up in general in Chapter III. Nevertheless it has been mentioned here, somewhat prematurely, because of the comparisons that have been made to pseudoreplication for ratio estimation.

3. Truncation Point

In Example 4.1 of Chapter I the general problem of estimation of the point of truncation of a distribution was discussed. The proper higher order estimators were introduced but were inappropriate for the uniform distribution being considered at that time, since $G^{(1)}(X_{(n)})$ was unbiased. Now we examine a situation in which the higher order generalized jackknife must be considered as having some potential.

Suppose we have a random sample $(X_1,\ldots,X_n)$ from a truncated exponential distribution. The general form of the density function is

$$f(x) = \frac{\sigma^{-1} \exp[-\frac{x-\alpha}{\sigma}]}{1 - \exp[-\frac{\theta-\alpha}{\sigma}]}, \quad \alpha < x < \theta$$

$$= 0 \quad , \quad \text{otherwise.}$$

The parameter α is known in most cases and we may, without loss of generality, take $\alpha = 0$ and $\sigma = 1$. Hence in this example we are considering a simple exponential which has been truncated at θ. The common distribution function of the X_i is

$$F(x) = \begin{cases} 0 & , \quad x < 0 \\ \dfrac{1 - e^{-x}}{1 - e^{-\theta}} & , \quad 0 \leq x \leq \theta \\ 1 & , \quad x > \theta \ . \end{cases}$$

If an estimate of the parameter θ is desired, then reliance upon the likelihood principle leads to the estimator, $\hat{\theta} = X_{(n)}$, the largest order statistic. The distribution function of $X_{(n)}$ is

$$P[X_{(n)} \leq x] = F_n(x) = \frac{(1 - e^{-x})^n}{(1 - e^{-\theta})^n} \quad .$$

Straightforward evaluation of the expected value of this estimator leads to

$$E[\hat{\theta}] = \int_0^{\theta} x \, dF_n(x)$$

$$= \theta - \int_0^{\theta} \frac{(1 - e^{-x})^n}{(1 - e^{-\theta})^n} dx$$

$$= \theta - (1 - e^{-\theta})^{-n} \int_0^{\theta} \sum_{r=0}^{n} (-1)^r \binom{n}{r} e^{-rx} dx$$

$$= \theta - \frac{\theta}{(1 - e^{-\theta})^n}$$

$$- (1 - e^{-\theta})^{-n} \sum_{r=1}^{n} (-1)^r \binom{n}{r} \int_0^{\theta} e^{-rx} dx \ .$$

Hence one expression for the bias of $\hat{\theta}$ is

$$\text{bias } [\hat{\theta}] = - \frac{1}{(1 - e^{-\theta})^n} \left[\theta - \sum_{r=1}^{n} (-1)^{r-1} \binom{n}{r} \frac{1 - e^{-r\theta}}{r} \right] .$$

This is of little value in constructing an estimator with less bias than $\hat{\theta}$. Alternatively, by recalling the method of Robson and Whitlock discussed in Chapter I we can derive a different expression as follows. Let

$$Y = \frac{1 - e^{-x}}{1 - e^{-\theta}} \quad .$$

Then

$$X = - \log [1 - Y(1 - e^{-\theta})] = H(Y) \quad ,$$

and

$$H'(Y) = \frac{(1 - e^{-\theta})}{[1 - Y(1 - e^{-\theta})]} \quad .$$

Consequently

$$H'(1) = \frac{1 - e^{-\theta}}{e^{-\theta}} = e^{\theta} - 1 \quad ,$$

and similarly

$$H''(1) = [e^{\theta} - 1]^2 \quad ,$$

and in general

$$H^{(k)}(1) = (k - 1)! \; [e^{\theta} - 1]^k \quad .$$

Therefore a factorial series expansion for the bias of the largest order statistic is given by

$$\text{bias } [X_{(n)}] = \sum_{k=1}^{\infty} (-1)^k \frac{n!}{(n + k)!} H^{(k)}(1)$$

$$= \sum_{k=1}^{\infty} (-1)^k \frac{n!(k-1)!}{(n+k)!} (e^{\theta} - 1)^k$$

$$= \frac{1}{n+1} \sum_{k=1}^{\infty} \frac{(-1)^k (e^{\theta} - 1)^k}{\binom{n+k}{k-1}}$$

$$= \frac{1}{n+1} \sum_{k=0}^{\infty} \frac{(1 - e^{\theta})^{k+1}}{\binom{k+n+1}{k}}$$

$$= \frac{(1 - e^{\theta})}{n+1} + \frac{1}{n+1} \sum_{k=1}^{\infty} \frac{(1 - e^{\theta})^{k+1}}{\binom{k+n+1}{k}} . \quad (3.1)$$

Hence by Theorem 3.4 of Chapter I, $G(\hat{\theta})$ L.O.B.E. $X_{(n)}$ if $R(n) = n/(n + 1)$, which we recall leads to the estimator

$$G(\hat{\theta}) = 2X_{(n)} - X_{(n-1)} .$$

We should note that in this example $J(\hat{\theta})$, given by

$$J(\hat{\theta}) = X_{(n)} + \frac{n-1}{n} [X_{(n)} - X_{(n-1)}] ,$$

is also L.O.B.E. $X_{(n)}$ and hence $G(\hat{\theta})$ and $J(\hat{\theta})$ should not be radically different. However, inspection of (3.1) suggests that in this problem $G(\hat{\theta})$ should have slightly better bias reduction properties.

The importance of eliminating additional terms in the expansion (3.1) can easily be seen when n is small and θ is large by inspection of that equation. The second-order generalized jackknife in this situation has previously been shown to be

$$G^{(2)}(\hat{\theta}) = 3X_{(n)} - 3X_{(n-1)} + X_{(n-2)} \quad .$$

A set of Monte Carlo runs employing 10,000 repetitions will illustrate the points mentioned above. First we have a situation in which the sample size is small (n = 5) and the point of truncation is large (θ = 4).

TABLE 3.1

Estimator	Average	Bias	MSE
$\hat{\theta} = X_{(n)}$	2.02	-1.98	4.62
$J(\hat{\theta})$	2.67	-1.33	3.44
$G(\hat{\theta})$	2.83	-1.17	3.37
$G^{(2)}(\hat{\theta})$	3.17	-0.83	4.91

Using either expression given above for the bias in $\hat{\theta}$, it can be shown that

$$E[\hat{\theta}] = \theta - \frac{\theta}{(1 - e^{-\theta})^n} + \frac{1}{(1 - e^{-\theta})^n} \sum_{r=1}^{n} \frac{(1 - e^{-\theta})^r}{r} . \tag{3.2}$$

When (3.2) is evaluated for n = 5 and θ = 4 we obtain $E[\hat{\theta}] = 2.0285$, so that it is safe to assume that in the table the second decimal place is within one or two units of the true value. Note that asymptotic unbiasedness is easy to verify from equation (3.2) since

$$\lim_{n\to\infty} E[\hat{\theta}] = \theta - \lim_{n\to\infty} \left[\frac{\theta - \sum_{r=1}^{n} (1 - e^{-\theta})^r / r}{(1 - e^{-\theta})^n} \right]$$

and by Theorem I.3.1

$$\lim_{n\to\infty} \left[\frac{\theta - \sum_{r=1}^{n} (1 - e^{-\theta})^r / r}{(1 - e^{-\theta})^n} \right]$$

$$= \lim_{n\to\infty} \left[\frac{-(1 - e^{-\theta})^n / n}{(1 - e^{-\theta})^n - (1 - e^{-\theta})^{n-1}} \right]$$

$$= 0 .$$

Nevertheless it is evident that the bias is quite serious for small sample sizes and that the jackknife affords a substantial improvement. The comparison of bias and MSE for $\hat{\theta}$ and $G^{(2)}(\hat{\theta})$ is striking in that the bias is significantly smaller for $G^{(2)}(\theta)$ with no appreciable increase in MSE. It is also worthy of note that both first-order jackknife estimators yield reductions in both bias and MSE over the MLE and furthermore

$$\text{bias } G(\hat{\theta}) < \text{bias } J(\hat{\theta}) < \text{bias } \hat{\theta}$$

and

$$\text{MSE } G(\hat{\theta}) < \text{MSE } J(\hat{\theta}) < \text{MSE } \hat{\theta} .$$

To see that the bias in $\hat{\theta}$ is nonnegligible even for moderate sample sizes and to point out again that the second-order estimator may not be desirable when the second term of the series expansion becomes small, let us examine the results in Table 3.2 for the case $n = 25$ and $\theta = 2$.

TABLE 3.2

Estimator	Average	Bias	MSE
$\hat{\theta}$	1.79	-0.21	0.071
$J(\hat{\theta})$	1.96	-0.04	0.061
$G(\hat{\theta})$	1.97	-0.03	0.063
$G^{(2)}(\hat{\theta})$	1.99	-0.01	0.162

The desirability of $G(\hat{\theta})$ over $\hat{\theta}$ is apparent. It must also be pointed out that even though $G^{(2)}(\hat{\theta})$ provides a further reduction in the bias of $\hat{\theta}$, the accompanying price in the variance is substantial. This increase in MSE is not as severe for larger values of θ. For instance, for the same sample size (n = 25) as that for Table 3.2 but $\theta = 4$, we have

$$\text{MSE } [G^{(2)}(\hat{\theta})] \cong \frac{3}{2} \text{ MSE } [\hat{\theta}] \quad .$$

Monte Carlo runs for all combinations of the sample sizes, n = 5, 10, 25, and the truncation parameters, $\theta = 2, 4$ exhibited a decrease in bias and MSE for both first-order jackknife estimators in all cases. Furthermore, in every case

$$\text{bias } [G(\hat{\theta})] < \text{bias } [J(\hat{\theta})] \quad .$$

4. Discriminant Analysis

When one is dealing with observations which are multivariate in nature, a problem which is frequently encountered is that of devising a rule for the classification of an individual multivariate observation as belonging to one of several different populations. A standard device, especially when there are only two distinguishable populations, is to form a linear combination of the elements of the vector and use the single value of this function to assign the observation to population 1 or population 2. When the observations $(X_1, X_2, \ldots, X_p)$

from each individual unit follow a multivariate normal distribution, the best linear combination has been studied and is relatively well-known. In this case the optimal linear compound is referred to as the linear discriminant function (LDF).

The assumptions which lead to this discrimination technique, which was originally introduced by Fisher [18], are that the two populations, π_1 and π_2, are multivariate normal with common covariance matrix Σ but different mean vectors $\underline{\mu}_1$ and $\underline{\mu}_2$. Under these assumptions and the additional assumptions that the costs of misclassification are equal and that the a priori probabilities of an individual observation arising from π_1 or π_2 are equal, the LDF can be shown to give rise to the Bayes rule for minimizing the sum of the two misclassification probabilities. If the unit associated with a random vector $\underline{X}$ is to be assigned to population π_1 or π_2, the LDF

$$Z = \underline{\alpha}'\underline{X}$$

is evaluated, and the rule states that the unit is classified as belonging to the population π_k whose mean $\underline{\alpha}'\underline{\mu}_k$ is closer to Z. The coefficients can be shown to be

$$\underline{\alpha} = \Sigma^{-1}(\underline{\mu}_1 - \underline{\mu}_2) ,$$

and the equivalent rule is as follows.

(1) Form

$$T = [\underline{x} - (1/2)(\underline{\mu}_1 + \underline{\mu}_2)]' \Sigma^{-1} (\underline{\mu}_1 - \underline{\mu}_2).$$

(2) Assign the unit to π_1 if $T > 0$,
assign the unit to π_2 if $T \leq 0$.

The quantity Z is the appropriate classification statistic in this case, and since the parameter values are usually unknown, a reasonable approach, which is ordinarily taken, is to replace Σ^{-1} and $\underline{\mu}_k$ by their maximum likelihood estimates, S^{-1} and $\overline{\underline{x}}_k$, respectively. In other words, the discriminant function used most often is

$$Z = \underline{a}'\underline{X}$$

where

$$\underline{a} = S^{-1}(\overline{\underline{X}}_1 - \overline{\underline{X}}_2) \quad .$$

An alternative way of developing this discriminant function is to determine the linear combination of observations having the greatest between-groups variation relative to its within-groups variation. This approach provides an intuitive justification for the use of the LDF under nonnormality. Otherwise there is nothing which necessarily recommends this statistic if the populations are not normal with common covariance matrix.

It is of interest in our current context to examine the expectation of $\underline{a}$ when normality is present. Suppose we have samples of size N_1 and N_2 from π_1 and π_2, respectively. The pooled unbiased estimator of the common covariance is

$$S = \frac{1}{N_1 + N_2 - 2} \left[\sum_{i=1}^{N_1} (\underline{X}_{1i} - \overline{\underline{X}}_1)^2 + \sum_{i=1}^{N_2} (\underline{X}_{2i} - \overline{\underline{X}}_2)^2 \right] ,$$

which is independent of both $\overline{\underline{X}}_1$ and $\overline{\underline{X}}_2$. Hence

$$E[\underline{a}] = E[S^{-1}(\overline{\underline{X}}_1 - \overline{\underline{X}}_2)]$$

$$= E[S^{-1}](\underline{\mu}_1 - \underline{\mu}_2),$$

and it has recently been shown by Das Gupta [14] that, for the p-dimensional Wishart matrix S with n degrees of freedom,

$$E[S^{-1}] = \frac{n}{n - p - 1} \Sigma^{-1} ,$$

where $n = N_1 + N_2 - 2$. Consequently

$$E[\underline{a}] = \frac{n}{n - p - 1} \underline{\alpha} ,$$

so that even under normality, $\underline{a}$ is a biased estimator of the true discriminant function coefficients. It is apparent, however, that the classification rule is not effected by scale changes and therefore the factor of $(n/(n - p - 1))$ is of no consequence. As a matter of fact many workers will alter the discriminant function, obtained by the foregoing procedure, to have a standard scale, i.e., the statistic which is often used is

$$Z = A(a_1x_1 + a_2x_2 + \ldots + a_px_p) + B$$

where A and B are chosen so that Z will equal 0 when $\underline{X} = \underline{\bar{X}}_1$ and Z will equal 1 when $\underline{X} = \underline{\bar{X}}_2$. The rule for classification becomes

$$\text{Assign the unit} \begin{cases} \text{to } \pi_1, \text{ if } Z < 1/2 \\ \text{to } \pi_2, \text{ if } Z \geq 1/2 \end{cases} .$$

If the multivariate populations are not normal it is questionable whether or not the jackknife coefficients, $J(\underline{a})$, actually represent an improvement in general. The subject of bias in the coefficients $\underline{a}$ is difficult to discuss when multivariate normality does not prevail. However, as we shall see in examples later on in Chapter III, the jackknifed discriminant function has, as a by-product, estimates of variability of the individual coefficients. In complex multivariate problems where little or no theory exists to give us exact tests of significance on coefficients, these variance estimates and resulting approximate confidence statements are reason enough to consider jackknifing the LDF when we suspect we are faced with nonnormality. An open task of some interest is an investigation of the penalty that is paid in the performance of the optimal linear discriminant function which has been jackknifed even though the data are multivariate normal.

Another facet of discriminant analysis, which has received attention because of the presence of bias, is the estimation of error rates of the LDF. This may be best explained by pointing out that once the discriminant function has been obtained the question usually arises as to the expected level of performance of the classification rule. Estimates of the two misclassification probabilities, $P(1|2)$ and $P(2|1)$, are needed, where $P(j|k)$ is the probability that a new observation of unknown origin will be classified as belonging to π_j given that it came from π_k. The techniques currently in use to estimate error rates may be divided into two classes; those which use sample data and those which use properties of the normal distribution. For a good comparison study of most of these methods see Lachenbruch and Mickey [25].

The methods which are dependent on normality are biased and if the degrees of freedom of the Wishart matrix S are small, they may seriously underestimate the probabilities of error. This is true under the assumption of multivariate normality and it is difficult to say how they perform as approximations for nonnormal situations in general. The empirical methods presented by Lachenbruch and Mickey probably represent the most promising approaches for nonnormal populations. The empirical method which Mosteller and Tukey [32] refer to as "cross-validation" is the same as the "U-method" of Lachenbruch and Mickey. The methods are summarized here to facilitate the discussion of this important application of the method which we have relied upon heavily to produce the second estimator for the jackknife statistic.

The H Method (Holdout). If we have a large number of samples, we could select a set from each group and compute a discriminant function from them, and then use the remainder of the observations to estimate the error rates. The number of misclassifications for each population would have binomial distributions with probabilities $P(1|2)$ and $P(2|1)$ and the straightforward estimates present no serious difficulties. After the estimates of $P(1|2)$ and $P(2|1)$ have been obtained, the discriminant function would be recomputed using the entire data set. There are several objections to this approach. First, large samples are typically not available in practice. Second, the discriminant function whose performance is evaluated is not the one that is eventually used. There may be an appreciable difference in the performance of the two rules. Third, the size of the holdout sample poses problems. If it is large, the estimate of performance which is obtained is good but that function may not perform well at all. If the holdout sample is small, the discriminant function will be relatively good but the estimator of its error probabilities unreliable. Finally, the method is not invariant, in the sense that different users would be likely to report different results from analyses of the same data. We shall not devote any further attention to this method.

The R Method (Resubstitution). This is simply a method in which the sample used to compute the discriminant function is reused directly to estimate error rates. Each observation is classified as belonging to π_1 or π_2 and the number of misclassifications is again treated as if it were binomially distributed. The fact that the standard error of this estimate of, say, $P(2|1)$, is not the usual binomial value $\sqrt{P(2|1)(1 - P(2|1))/N_1}$ does not appear to be the major difficulty. The method has been found to be very misleading. Even for moderate sample sizes the estimates of performance are overly optimistic. The bias here is serious since the rule is judged on the same data set from which it was tailored.

The U Method. This procedure makes use of all observations as does the R method without such serious biases in the estimated error rates. Let us denote our original discriminant function evaluated at a new observation vector $\underline{x}$ by

$$Z = D(\underline{x}) = \underline{a}'\underline{x} = \underline{x}'S^{-1}(\overline{\underline{x}}_1 - \overline{\underline{x}}_2) ,$$

and the discriminant function computed from all the observations except $\underline{x}_j$ by D_j. The function D_j may now be used to classify $\underline{x}_j$. When this has been done for each $\underline{x}_j$ in, say, π_1 then the estimate of misclassification probability $P(2|1)$ is obtained by dividing the total number misclassified by N_1, the number of samples which we have from π_1.

To do this requires $N_1 + N_2$ discriminant functions, but an identity due to Bartlett [5] allows us to avoid an equal number of matrix inversions. Bartlett showed that if A and B are square nonsingular matrices and $\underline{u}$ and $\underline{v}$ are column vectors such that

$$B = A + \underline{u}\,\underline{v}' ,$$

then

$$B^{-1} = A^{-1} - [A^{-1}\underline{u}\,\underline{v}'\,A^{-1}/(1 + \underline{v}'A^{-1}\underline{u})] .$$

Therefore let us define

$$\underline{U}_j = \underline{x}_j - \bar{\underline{x}}_k ,$$

when $\underline{x}_j$ belongs to $\pi_k (k = 1,2)$. Then the sample pooled covariance matrix computed with $\underline{x}_j$ removed from π_k may be written

$$S_{(j)} = S - c_k\,\underline{U}_j\,\underline{U}_j' ,$$

where

$$c_k = \frac{N_k}{(N_k - 1)(N_1 + N_2 - 2)} .$$

Hence it can be shown that

$$S^{-1}_{(j)} = \frac{N_1 + N_2 - 3}{N_1 + N_2 - 2} \left[S^{-1} + \frac{c_k S^{-1} \underline{U}_j \, \underline{U}'_j \, S^{-1}}{1 - c_k \, \underline{U}'_j \, S^{-1} \underline{U}_j} \right] ,$$

and the procedure requires only one explicit inversion. This identity is also useful in reducing the labor required to form the second estimator for the jackknife.

Next it can be shown that to adjust the difference of sample means, $\underline{d} = \bar{\underline{x}}_1 - \bar{\underline{x}}_2$, for the omission of $\underline{x}_j$ from π_k we may use

$$\underline{d}_{(j)} = \underline{d} + \frac{(-1)^k \, \underline{U}_j}{N_k - 1} .$$

Hence to use the discriminant function D_j to classify $\underline{x}_j$ we compute

$$D_j(\underline{x}_j) = \underline{x}'_j \, S^{-1}_{(j)} \, \underline{d}_{(j)} .$$

There is correlation between $D_i(\underline{x}_i)$ and $D_j(\underline{x}_j)$,

$i \neq j$, but this is small in most cases (see [25]).

We shall now turn to a specific example in discriminant analysis. The U method will be seen to be related to the jackknife approach presented by Mosteller and Tukey [32]. When these latter authors presented a procedure for "cross-validation," the U method appeared in print at about the same time and there appears to be some confusion concerning their equivalence. In a commentary on the paper of Lachenbruch and Mickey by Cochran [12] the two procedures are referenced as distinct approaches which in fact they are not.

This example deals with determining the authorship of papers from the rates of appearance of certain high-frequency words. Alexander Hamilton and James Madison both wrote numerous politcial papers during the same period of American history. Extensive studies of this general problem of authorship of various Federalist papers have been made elsewhere and this small study first given in [32] is merely illustrative. The goal here is to discriminate between the papers of Hamilton and those of Madison on the basis of five words which they both used with a high frequency. This approach was selected without regard to prior study of its discriminatory power but to represent a moderate level of complexity. In describing the data used for this example Mosteller and Tukey [32] state that:

> We chose 11 papers known to have been written by each author, mainly from among

the Federalist papers. These particular 22 papers were chosen because among the 100 or so papers we had available, their lengths were nearest to 2500 words, running from about 2200 to about 2800. For some purposes, it would have been better to have chosen randomly. For convenience in applying the jackknife, each Hamilton paper was randomly paired with a Madison paper. Perhaps we could have paired them more meaningfully by order of publication, but we did not. The number $k = 11$ was chosen partly because it is one more than the round number 10, and we frequently need to multiply or divide by $k - 1$. Also 10 is only about twice as big as 5, the number of variables chosen for analysis, and one of our purposes is to illustrate the variability that may occur in a study of several weakly discriminating variables when we have only a modest set of data available for establishing the technique of making distinctions.

The observations recorded for each paper consist of the rates per thousand words (denoted x_i) for the words

"and"	(x_1)
"in"	(x_2)
"of"	(x_3)

"the" (x_4)

"to" (x_5)

Table 4.1 contains the two multivariate samples and in our example $p = 5$, $N_1 = N_2 = 11$. Table 4.2 gives the sums of squares and cross-products of deviations in each group and pooled. The difference in sample mean vectors is also given. Only the lower triangle of the symmetric 5×5 matrices are presented.

Rather than computing the discriminant function coefficients as previously described, let us take an equivalent approach, namely, regression of dummy variables, indicating π_1 or π_2, on the observations $\{\underline{x}_j\}$ (see Kshirsagar [24]). We take this alternative approach since this was the method utilized originally by Mosteller and Tukey for this problem. In this example it was convenient to assign each Hamilton paper a value of Z equal to 1 and each Madison paper a Z value of 0, and then fit a multiple regression function,

$$D(\underline{x}) = Z = A(a_1x_1 + a_2x_2 + a_3x_3 + a_4x_4 + a_5x_5) + B .$$

In doing so, the free coefficients A and B are automatically selected so that

$$\frac{1}{N_1} \sum_{j=1}^{N_1} D(\underline{x}_j) = 1 ,$$

when the $\underline{x}_j$ come from π_1 (Hamilton), and

TABLE 4.1

	Observation j	and 1 x_1	in 2 x_2	of 3 x_3	the 4 x_4	to 5 x_5	
		Word i					
Hamilton (π_1)	1	16.1	35.3	63.9	98.3	38.4	
	2	32.2	24.5	78.2	110.0	31.4	
	3	24.3	23.5	64.7	90.8	42.3	
	4	18.0	27.2	59.6	86.8	35.9	
	5	20.6	26.9	61.4	83.6	39.5	
	6	21.8	17.4	73.1	90.4	35.6	
	7	27.9	23.1	61.9	85.4	41.3	
	8	28.5	26.1	71.3	74.5	33.3	
	9	28.9	20.9	56.9	82.7	44.9	
	10	21.3	25.0	60.4	82.2	47.7	
	11	18.5	30.7	72.7	109.3	36.6	
	Sum	(258.1,	280.6,	724.1,	994.0,	426.9)	$= N_1\bar{\underline{x}}_1'$
Madison (π_2)	1	31.6	19.9	54.8	93.8	38.6	
	2	37.3	23.3	56.8	84.2	31.0	
	3	21.2	17.5	58.2	97.6	39.9	
	4	27.9	19.1	55.8	93.1	33.5	
	5	40.7	9.3	59.0	71.5	33.6	
	6	24.4	27.9	60.0	115.3	34.8	
	7	27.7	17.7	61.1	115.3	32.7	
	8	28.1	22.3	57.0	110.9	29.7	
	9	30.6	23.6	68.3	118.6	23.2	
	10	33.9	21.8	64.9	93.7	33.6	
	11	23.3	31.4	34.8	94.3	49.6	
	Sum	(326.7,	233.8,	630.7,	1088.3,	380.2)	$= N_2\bar{\underline{x}}_2'$

TABLE 4.2

Hamilton

	275.985				
	-139.756	226.069			
$(N_1 - 1)S_1 =$	95.181	-12.473	471.102		
	-67.655	181.644	455.111	1267.465	
	-31.396	-34.801	-260.843	-227.696	244.069

Madison

	351.920				
	-173.050	334.287			
$(N_2 - 1)S_2 =$	169.590	-212.312	719.265		
	-514.910	381.708	364.715	2146.385	
	-173.710	109.502	-479.175	-315.635	442.865

	637.905				
	-312.806	560.356			
$(N_1+N_2-2)S =$	264.771	-224.785	1190.367		
	-582.565	563.352	819.826	3413.850	
	-205.106	74.701	-740.018	-543.331	686.934

Hamilton sum minus Madison sum

$(11\underline{d}') = (-68.6 ,\ 46.8 ,\ 93.4 ,\ -94.3 ,\ 46.7)$

$$\frac{1}{N_2} \sum_{j=1}^{N_2} D(\underline{x}_j) = 0 ,$$

when the $\underline{x}_j$ come from π_2 (Madison). Since we have equal group sizes and assume that no *a priori* information would be at hand if we were to employ the decision rule on an unknown paper, we shall rule that a discriminant score greater than 1/2 is a Hamilton indicator and it shall be classified as a Madison paper otherwise.

Table 4.3 gives the coefficients for the five variables and the constant term for the discriminant function D and for each function D_j obtained by dropping the jth pair of papers from the sample.

We will use the D_j to classify the jth Hamilton paper, the jth Madison paper, and obtain estimates of the error probabilities by the U method. Before proceeding to this let us examine the use of the D_j to form the jackknifed discriminant function D*. First notice that we are jackknifing the function and not values of the function. Since the jackknife is a linear operator (a fact we have not mentioned previously), we have

$$J[D(\underline{x})] = J(a_1)\ x_1 + \ldots + J(a_5)x_5 + J(B) \quad .$$

Consequently, in this specific example the function is jackknifed by applying Quenouille's method coefficient by coefficient. For example to obtain the third coefficient in the "pseudo-discriminant-function," $J_4(D)$, associated with deleting the fourth matched pair, compute

$$11(\text{coefficient of } x_3 \text{ in } D) - 10(\text{coefficient of } x_3 \text{ in } D_4) .$$

TABLE 4.3

	Coefficient of: x_1	x_2	x_3	x_4	x_5	Constant term
D	-0.01902	0.02851	0.05264	-0.01642	0.04056	-2.83668
D_1	-0.01904	0.03032	0.05295	-0.01660	0.04120	-2.89257
D_2	-0.02747	0.02559	0.04532	-0.01964	0.03163	-1.47660
D_3	-0.02884	0.01467	0.04928	-0.01479	0.03962	-2.14043
D_4	-0.00716	0.02874	0.05631	-0.01248	0.05243	-4.21632
D_5	-0.01790	0.02789	0.05348	-0.01642	0.04172	-2.94733
D_6	-0.01695	0.03151	0.05182	-0.01455	0.04145	-3.10996
D_7	-0.02053	0.03063	0.05166	-0.01757	0.03681	-2.55988
D_8	-0.01648	0.03338	0.05660	-0.01979	0.04184	-2.99265
D_9	-0.02350	0.02910	0.05002	-0.01670	0.03074	-2.20700
D_{10}	-0.01093	0.03047	0.05406	-0.01559	0.04523	-3.40123
D_{11}	-0.01983	0.02953	0.05521	-0.01616	0.04188	-3.06431

Thus, from Table 4.3,

$$J_4(D) = 11(0,0526442) - 10(0.0563169) = 0.015917 \quad .$$

Table 4.4 presents the coefficients of the "pseudo-functions," $J_j(D)$, and of the jackknifed discriminant function $D^* = J(D)$. The example result just obtained may be located in the fourth row of coefficients in the third column.

TABLE 4.4

	Coefficient of: x_1	x_2	x_3	x_4	x_5	Constant term
$J_1(D)$	-0.01886	0.01041	0.04956	-0.01463	0.03422	-2.27783
$J_2(D)$	0.06548	0.05770	0.12584	0.01582	0.12988	-16.43747
$J_3(D)$	0.07914	0.16695	0.08621	-0.03268	0.04996	-9.79921
$J_4(D)$	-0.13765	0.02623	0.01591	-0.05584	-0.07806	10.95966
$J_5(D)$	-0.03020	0.03477	0.04425	-0.01639	0.02902	-1.73025
$J_6(D)$	-0.03967	-0.00149	0.06081	-0.03513	0.03166	-0.10390
$J_7(D)$	-0.00389	0.00736	0.06242	-0.00490	0.07810	-5.60468
$J_8(D)$	-0.04443	-0.02015	0.01308	0.01731	0.02776	-1.27696
$J_9(D)$	0.02582	0.02262	0.07885	-0.01364	0.13878	-9.13354
$J_{10}(D)$	-0.09988	0.00897	0.03843	-0.02470	-0.00606	2.80880
$J_{11}(D)$	-0.01090	0.01831	0.02697	-0.01905	0.02736	-0.56044
D^*	-0.01955	0.03015	0.05476	-0.01671	0.04205	-3.01416

We can use the R method to investigate the misclassification probabilities of the original rule which utilizes D and of the rule which uses D*. Table 4.5 gives the values of D and D* obtained by resubstituting each of the 22 observation vectors from the Hamilton and Madison papers into the functions.

TABLE 4.5

Matched pair of papers	Function			
	D		D*	
	Author		Author	
	H	M	H	M
1	1.170	0.039	1.206	0.024
2	0.833	-0.017	0.859	-0.033
3	1.001	0.338	1.023	0.333
4	0.764	-0.055	0.777	-0.075
5	1.000	-0.051	1.020	-0.080
6	1.052	0.171	1.073	0.172
7	0.822	-0.209	0.836	-0.227
8	1.246	-0.351	1.275	-0.374
9	0.668	-0.157	0.673	-0.167
10	1.235	0.380	1.263	0.381
11	1.203	-0.089	1.243	-0.107
Average	0.999	0.000	1.023	-0.014

There is a marked similarity between the evaluations of the two functions on the 22 papers. Furthermore all of the papers are seen to be correctly classified by our rule which uses the value 1/2 as the partition. These are the potentially misleading estimates for which the U method offers some improvement.

Recall the 11 functions D_j given in Table 4.3. When these functions are evaluated on the papers omitted from the computation of their coefficients, we obtain the 22 values, $D_j(\underline{x}_j)$, given in Table 4.6. Note that this approach yields an estimate of the misclassification probability for Madison papers of 1/11.

TABLE 4.6

Paper Pair j	D_j applied to H_j	D_j applied to M_j
1	1.205	-0.044
2	0.642	-0.004
3	1.025	0.510†
4	0.592	-0.130
5	0.993	-0.034
6	1.018	0.230
7	0.792	-0.253
8	1.363	-0.438
9	0.567	-0.091
10	1.269	0.459
11	1.256	-0.124
Average	0.975	+0.015

†This paper misclassified.

It must be stressed that this technique provides us with an estimate of the performance of the original rule based on D and not for D*. If the intent is to use D* because of the accompanying estimate of stability (to be discussed in Chapter III), then to get estimated error rates, the U method must be applied to each of the "pseudo-discriminants" of Table 4.4. This seems formidable at first but actually it should be no more so than what has already been done here with the aid of the computer and a labor saving identity. The procedure requires the calculation of discriminant functions on all subsets of size n - 2, which has the flavor of the second-order jackknife, but it is actually a combination of a first-order jackknife and the U method. This similarity exists between the U method and a single application of the jack-knife, accounting for the impression that users of the U method are jackknifing. It is apparent that the estimated error rates have not been jackknifed and the U method may be used without proceeding to jackknife the discriminant function D.

Mosteller and Tukey carried out the estimation of error rates for D* and by averaging the observed error rates for the "pseudodiscriminants," we obtain equal probabilities of misclassification of $17/110 \cong 15.4\%$. This is very likely a more realistic estimate of the quality of performance of a classification rule based on this much data of this nature.

An example of the U method approach in a problem of greater complexity and magnitude may be found in

Dempster [16] (pp. 255-260). The analysis is somewhat advanced but the graphical presentations are instructive.

CHAPTER III
ASYMPTOTIC DISTRIBUTIONS

1. Introduction

In Chapters I and II we have only considered the jackknife as a method for reducing bias or as an aid in producing unbiased estimators. However, although this was the original motivation for the jackknife as a statistical tool, it is the asymptotic properties of the jackknife that make it a useful general tool in data analysis. This facet of $J(\hat{\theta})$ was first suggested by Tukey in [46] where he proposed that the pseudovalues $J_i(\hat{\theta})$ could be treated as independent identically distributed random variables, and hence the jackknife could be used to obtain approximate confidence intervals. To be more specific, suppose one has a random sample of size n and an estimator $\hat{\theta}$ for θ. Then by splitting the sample into N groups of size M and forming the jackknife one obtains in the process the pseudovalues $J_i(\hat{\theta})$, i.e., the random variables

$$J_i(\hat{\theta}) = N\hat{\theta} - (N - 1)\hat{\theta}^i \quad . \tag{1.1}$$

Now Tukey's conjecture was that the $J_i(\hat{\theta})$ are approximately independent and hence one would expect the quantity

$$\frac{(J(\hat{\theta}) - \theta)\sqrt{N(N-1)}}{\sqrt{\sum_{i=1}^{N} (J_i(\hat{\theta}) - J(\hat{\theta}))^2}} \tag{1.2}$$

to be approximately distributed as a t random variable. Thus one could obtain an approximate confidence interval for θ. The procedure suggested by this conjecture was subsequently used by Tukey and others with quite satisfactory results. The theoretical support of such a procedure is the subject of the next section. Unless specified to the contrary we will take $M = 1$.

2. Asymptotic Theorems

The first theorem we shall consider is an extension of a result due to Miller [30]. The interest in this theorem, as well as the theorems following it, stems not only from Tukey's conjecture but from the fact that statisticians often find themselves using estimators of the form $f(\hat{\phi})$ for $f(\phi)$, where $\hat{\phi}$ is an unbiased estimator of ϕ. In this situation $f(\hat{\phi})$ will usually be biased and hence a perfect candidate for the jackknife.

In the remaining theorems of this chapter we will denote the pseudovalues associated with $G(\hat{\theta})$ by $G_i(\hat{\theta})$, i.e.,

$$G_i(\hat{\theta}) = \frac{\hat{\theta} - R(n)\hat{\theta}^i}{1 - R(n)} \quad . \tag{2.1}$$

<u>THEOREM 2.1</u>. Let $X_1, X_2, \ldots, X_n$ be a random sample from a distribution with mean μ and finite variance σ^2. Let

$$\theta = f(\mu) \quad ,$$

$$\overline{X}_n = \frac{1}{n} \sum_{i=1}^{n} X_i \quad ,$$

and

$$\hat{\theta} = f(\overline{X}_n) \quad , \tag{2.2}$$

where f is a real-valued function, defined on the real line, which possesses a bounded second derivative in a neighborhood of μ.

Further suppose

$$\alpha = \lim_{n\to\infty} \frac{R(n)}{(1 - R(n))(n - 1)} \neq \pm \infty, 0 \ . \tag{2.3}$$

Then the random variable

$$\frac{\alpha(G(\hat{\theta}) - \theta)\sqrt{n(n-1)}}{\sqrt{\sum_{i=1}^{n} (G_i(\hat{\theta}) - G(\hat{\theta}))^2}} \underset{\mathcal{L}}{\rightarrow} N(0,1), \tag{2.4}$$

as $n \to \infty$, i.e., the random variable in equation (2.4) is asymptotically distributed as a normal random variable with mean zero and variance one.

Before proving the above result let us first give two additional theorems.

THEOREM 2.2. Given the conditions of Theorem 2.1 the random variable $(G(\hat{\theta}) - \theta)\sqrt{n}$ is asymptotically distributed $N(0,\sigma_1^2)$, as $n \to \infty$, where

$$\sigma_1^2 = \sigma^2[f'(\mu)]^2 \quad . \tag{2.5}$$

Proof: We shall first expand f in a Taylor series about μ. To do this we need to show that values of $\bar{X}_n$ and $\bar{X}_{n-1}^i$ simultaneously fall within an arbitrarily small interval about μ with probability approaching unity. Before expanding f we therefore proceed as follows. Let $I = (\mu - 3\Delta, \mu + 3\Delta)$, $\Delta > 0$, be any neighborhood of μ in which f" is bounded. Then

$$P[\max\{\frac{|X_1|}{n}, \frac{|X_2|}{n}, \ldots, \frac{|X_n|}{n}\} > \Delta]$$

$$= 1 - [F_{|X|}(n\Delta)]^n \tag{2.6}$$

where $F_{|X|}$ is the distribution function of $|X|$. Now to find the limit as $n \to \infty$ in (2.6) consider

$$\lim_{t\to\infty} \log\,[1 - (1 - F_{|X|}(t))]^t$$

$$= \lim_{t\to\infty} t[-(1 - F_{|X|}(t)$$

$$+ 0\,(1 - F_{|X|}(t))^2] \quad , \tag{2.7}$$

where 0 is the order symbol. But for $t > 0$,

$$t(1 - F_{|X|}(t)) = t \int_t^\infty dF_{|X|}(u) \leq \int_t^\infty u \, dF_{|X|}(u) \; .$$

Since $\sigma^2 < \infty$, $E[|X|] < \infty$ and hence the latter integral exists. Therefore

$$\lim_{t\to\infty} t \, (1 - F_{|X|}(t)) \;=\; 0 \quad ,$$

which implies that $\lim_{t\to\infty} t\{0[(1 - F_{|X|}(t))^2]\} = 0$ as well. Thus $[F_{|X|}(n\Delta)]^n \to 1$ and consequently

$$\lim_{n\to\infty} P[\max \{ \frac{|X_1|}{n} , \frac{|X_2|}{n} , \ldots , \frac{|X_n|}{n} \} > \Delta] = 0 \; . \tag{2.8}$$

Now

$$\bar{X}^i_{n-1} = \frac{1}{n - 1} \sum_{\substack{j=1 \\ j\neq i}}^{n} X_j \quad ,$$

or alternatively

$$\bar{X}^i_{n-1} = \frac{n}{n - 1} \bar{X}_n - \frac{1}{n - 1} X_i \; ,$$

and therefore

$$|\overline{X}^{i}_{n-1} - \mu| \leq \frac{n}{n-1} |\overline{X}_n - \frac{n-1}{n} \mu| + \frac{1}{n-1} |X_i|$$

$$\leq \frac{n}{n-1} [|\overline{X}_n - \mu| + \frac{|\mu|}{n} + \frac{X_i}{n}] .$$

Then letting $I_1 = (-3\Delta, 3\Delta)$ we have

$$\lim_{n\to\infty} P[(\overline{X}_n - \mu) \varepsilon I_1, (\overline{X}^{1}_{n-1} - \mu) \varepsilon I_1, \ldots (\overline{X}^{n}_{n-1} - \mu) \varepsilon I_1]$$

$$= \lim_{n\to\infty} P[\max |\overline{X}^{i}_{n-1} - \mu| < 3\Delta]$$

$$\geq \lim_{n\to\infty} P[\frac{n}{n-1} (|\overline{X}_n - \mu| + \frac{|\mu|}{n} + \frac{1}{n} \max \{|X_i|\}) < 3\Delta]$$

$$= \lim_{n\to\infty} P[|\overline{X}_n - \mu| + \frac{1}{n} \max\{|X_i|\} < 3\Delta - \frac{1}{n} (3\Delta + |\mu|)]$$

$$\geq \lim_{n\to\infty} P[|\overline{X}_n - \mu| + \frac{1}{n} \max \{|X_i|\} < 2\Delta]$$

$$\geq \lim_{n\to\infty} P[|\overline{X}_n - \mu| < \Delta, \frac{1}{n} \max \{|X_i|\} < \Delta]$$

$$= \lim_{n\to\infty} P[\frac{1}{n} \max \{|X_i|\} < \Delta]$$

$$\cdot P[|\overline{X}_n - \mu| < \Delta \,|\frac{1}{n} \max \{|X_i|\} < \Delta]$$

$$= 1$$

since $\overline{X}_n \to \mu$ in probability. It follows then that

$$\lim_{n\to\infty} P[\overline{X}_n \in I, \overline{X}^1_{n-1} \in I, \ldots, \overline{X}^n_{n-1} \in I] = 1. \quad (2.9)$$

From (2.9), we have the probability that the equations below hold approaches one as $n \to \infty$. That is,

$$f(\overline{X}^i_{n-1}) = f(\overline{X}_n) + (\overline{X}^i_{n-1} - \overline{X}_n)f'(\overline{X}_n)$$

$$+ \frac{(\overline{X}^i_{n-1} - \overline{X}_n)^2}{2} f''(Y_i) \quad ,$$

where $|\overline{X}_n - Y_i| < |\overline{X}_n - \overline{X}^i_{n-1}|$, or, more correctly stated, Y_i is a random variable such that for given values of $\overline{X}_n$ and $\overline{X}^i_{n-1}$, the random variable Y_i takes on values satisfying the inequality

$$|\bar{x}_n - y_i| < |\bar{x}_n - \bar{x}^i_{n-1}|$$

with probability 1. Then

$$G(\hat{\theta}) = \left\{ f(\bar{X}_n) - R(n) \frac{1}{n} \sum_{i=1}^{n} [f(\bar{X}_n) + (\bar{X}^i_{n-1} - \bar{X}_n) f'(\bar{X}_n) + \frac{(\bar{X}^i_{n-1} - \bar{X}_n)^2}{2} f''(Y_i)] \right\} \bigg/ [1 - R(n)]$$

$$= f(\bar{X}_n) - \frac{R(n)}{2(1 - R(n))n} \sum_{i=1}^{n} (\bar{X}^i_{n-1} - \bar{X}_n)^2 f''(Y_i) .$$

Thus

$$\sqrt{n}(G(\hat{\theta}) - \theta) = \sqrt{n}[f(\bar{X}_n) - f(\mu)] - \frac{R(n)}{2(1 - R(n))\sqrt{n}} \sum_{i=1}^{n} (\bar{X}^i_{n-1} - \bar{X}_n)^2 f''(Y_i)$$

$$= \sqrt{n}[f(\bar{X}_n) - f(\mu)] - \frac{R(n)}{2(1 - R(n))\sqrt{n}(n - 1)^2} \sum_{i=1}^{n} (X_i - \bar{X}_n)^2 f''(Y_i) .$$

Now again with probability approaching 1 as $n \to \infty$ we can write

$$f(\overline{X}_n) = f(\mu) + (\overline{X}_n - \mu) f'(Y) \quad ,$$

or

$$\sqrt{n}[f(\overline{X}_n) - f(\mu)] = \sqrt{n}(\overline{X}_n - \mu)\ f'(Y) \quad ,$$

where $|Y - \mu| < |\overline{X}_n - \mu|$. Furthermore since

$$f'(Y) \underset{p}{\to} f'(\mu) \quad \text{as } n \to \infty \quad ,$$

then $\sqrt{n}[f(\overline{X}_n) - f(\mu)]$ approaches a $N(0,\sigma_1^2)$ random variable in probability and thus is asymptotically $N(0,\sigma_1^2)$ (recall equation (2.5)). Now let $|f''(t)| < M$ for all $t \in I$. Then for values of $\overline{X}_n$, $\overline{X}^i_{n-1} \in I$ we have

$$\left| \sum_{i=1}^{n} (x_i - \overline{x}_n)^2 f''(y_i) \right|$$
$$< M \sum_{i=1}^{n} (x_i - \overline{x}_n)^2 \quad .$$

But since the sample variance is consistent and α is finite (see equation (2.3)) we have

$$\frac{1}{2\sqrt{n}}\left[\frac{R(n)}{(1 - R(n))(n - 1)}\right]\left[\frac{1}{n - 1}\sum_{i=1}^{n}(X_i-\overline{X}_n)^2 f''(Y_i)\right]$$

$$\underset{p}{\rightarrow} \quad 0 \quad ,$$

and hence $\sqrt{n}(G(\hat{\theta}) - \theta)$ is asymptotically $N(0,\sigma_1^2)$. This completes the proof of Theorem 2.2.

The proof of the next theorem is much the same as the proof above and hence will only be sketched. For a proof when $R(n) = (n - 1)/n$ see [30].

<u>THEOREM 2.3</u>. Let $\{X_i\}$ be a sequence of independent identically distributed random variables with mean μ, and finite variance σ^2. Let f be a function with a continuous first derivative in a neighborhood of μ. Then as $n \rightarrow \infty$, $S_G^2 \underset{p}{\rightarrow} \sigma^2[f'(\mu)]^2\alpha^2 = \sigma_1^2\alpha^2$,

where

$$S_G^2 = \frac{1}{n - 1}\sum_{i=1}^{n}[G_i(\hat{\theta}) - G(\hat{\theta})]^2 \quad .$$

Proof: In the same manner as the previous proof we can write the following equation with probability approaching 1 as $n \rightarrow \infty$:

$$f(\overline{X}_{n-1}^i) = f(\overline{X}_n) + (\overline{X}_{n-1}^i - \overline{X}_n)f'(Y_i) \quad ,$$

where

$$\left|\overline{X}_n - Y_i\right| < \left|\overline{X}_n - \overline{X}^i_{n-1}\right| .$$

Then

$$S_G^2 = \frac{1}{(1 - R(n))^2 (n - 1)} \sum_{i=1}^{n} [f(\overline{X}_n) - R(n) f(\overline{X}^i_n)$$

$$- f(\overline{X}_n) + \frac{R(n)}{n} \sum_{j=1}^{n} f(\overline{X}^j_{n-1})]^2$$

$$= \frac{[R(n)]^2}{(1 - R(n))^2} \cdot \frac{1}{n - 1} \sum_{i=1}^{n} [f(\overline{X}^i_n)$$

$$- \frac{1}{n} \sum_{j=1}^{n} f(\overline{X}^j_{n-1})]^2$$

$$= \left(\frac{R(n)}{1 - R(n)}\right)^2 \left(\frac{1}{n - 1}\right) \sum_{i=1}^{n} [(\overline{X}^i_{n-1} - \overline{X}_n) f'(Y_i)$$

$$- \frac{1}{n} \sum_{j=1}^{n} (\overline{X}^j_{n-1} - \overline{X}_n) f'(Y_j)]^2$$

$$= \left(\frac{R(n)}{1 - R(n)}\right)^2 \left(\frac{1}{n - 1}\right) \sum_{i=1}^{n} [- \frac{1}{n - 1} (X_i - \overline{X}_n) f'(Y_i)$$

$$+ \frac{1}{n(n - 1)} \sum_{j=1}^{n} (X_j - \overline{X}_n) f'(Y_j)]^2$$

$$= \left(\frac{R(n)}{1 - R(n)}\right)^2 \left(\frac{1}{n - 1}\right)^3 \sum_{i=1}^{n} [(X_i - \overline{X}_n) f'(\mu)$$

$$+ (X_i - \overline{X}_n)(f'(Y_i) - f'(\mu))$$

$$- \frac{1}{n} \sum_{j=1}^{n} (X_j - \overline{X}_n)(f'(Y_j) - f'(\mu))]^2$$

$$= \frac{1}{(n-1)^2} \left(\frac{R(n)}{1 - R(n)}\right)^2 [f'(\mu)]^2 \left[\frac{1}{n-1} \sum_{i=1}^{n} (X_i - \overline{X}_n)^2\right] + \text{(terms which approach zero in probability)} \quad . \qquad (2.10)$$

The result then follows immediately from (2.10). We are now in a position to prove Theorem 2.1.

Proof of Theorem 2.1: By Theorem 2.2

$$\sqrt{n}\ (G(\hat{\theta}) - \theta) \underset{\mathcal{L}}{\to} N(0,\sigma_1^2)$$

and hence

$$\alpha\sqrt{n}\ (G(\hat{\theta}) - \theta) \underset{\mathcal{L}}{\to} N(0,\alpha^2\sigma_1^2) \ .$$

By Theorem 2.3, $S_G^2 \underset{P}{\to} \sigma_1^2\alpha^2$ and hence

$$\frac{\sqrt{n}\,(G(\hat{\theta}) - \theta)}{\sqrt{\dfrac{1}{n-1}\sum_{i=1}^{n}[G_i(\hat{\theta}) - G(\hat{\theta})]^2}} \underset{\mathcal{L}}{\rightarrow} N(0,1)$$

as was to be shown.

Theorem 2.1 can be generalized to functions of U-statistics. We will simply state that generalization here. A proof of the result for $R(n) = (n - 1)/n$ can be found in [3] and the proof is essentially no different for a general value of $R(n)$.

THEOREM 2.4. Let $X_1, X_2, \ldots, X_n$ be a random sample from a distribution with parameter β. Let U_n be a U-statistic such that

$$U_n(X_1, X_2, \ldots, X_n) = \binom{n}{m}^{-1} \sum_{C} g(X_{i_1}, X_{i_2}, \ldots, X_{i_m})$$

where C indicates that the summation is over all combinations $(i_1, i_2, \ldots, i_m)$ of m integers chosen from $(1, 2, \ldots, n)$,

$$E[g(X_1, X_2, \ldots, X_m)] = \beta ,$$

and

$$E[(g(X_1, X_2, \ldots, X_m))^2] < \infty$$

where $g(X_1, X_2, \ldots, X_m)$ is a symmetric estimator and

$m \leq n$. Further let

$$\theta = f(\beta)$$

and

$$\hat{\theta} = f(U_n)$$

where f is a real-valued function defined on the real line with a bounded second derivative in a neighborhood of β. Then the random variable

$$\frac{\alpha(G(\hat{\theta}) - \theta)\sqrt{n(n-1)}}{\sqrt{\sum_{i=1}^{n} [G_i(\hat{\theta}) - G(\hat{\theta})]^2}} \underset{\mathcal{L}}{\rightarrow} N(0,1) \tag{2.11}$$

as $n \to \infty$.

Note that if

$$R(n) = \frac{n-1}{n},$$

then

$$\frac{R(n)}{1 - R(n)} \cdot \frac{1}{n-1} \equiv 1, \tag{2.12}$$

and hence when

$$G(\hat{\theta}) = J(\hat{\theta})$$

the multiplier α is unity. As we have seen in Chapter II, when some information about $B(n,\theta)$, the bias in $\hat{\theta}$, can reasonably be assumed we may desire to select $R(n)$ differently than $(n - 1)/n$. For example, if sufficient information is known about our model to predict $B(n,\theta) = O(n^{-1/2})$ we would probably take

$$R(n) = \sqrt{\frac{n-1}{n}} .$$

However, with this choice of $R(n)$ we note

$$\lim_{n\to\infty} \frac{R(n)}{1 - R(n)} \cdot \frac{1}{n-1} = \lim_{n\to\infty} \frac{1}{1 - \sqrt{\frac{n-1}{n}}} \cdot \frac{1}{n-1}$$

$$= \lim_{n\to\infty} \frac{\sqrt{n} + \sqrt{n-1}}{\sqrt{n}} = 2 ,$$

and hence there may be instances where $\alpha \neq 1$. A theorem of interest along this line is the following.

<u>THEOREM 2.5</u>. Let $B(n,\theta) \neq 0$ and assume there exists a $p > 0$ such that

$$\lim_{n\to\infty} n^p B(n,\theta) = C(\theta) \neq 0 \text{ or } \pm\infty.$$

Then if

$$R(n) = ((n - 1)/n)^p ,$$

$G(\hat{\theta})$ is L.O.B.E. $\hat{\theta}$ and if the conditions of Theorem 2.4 hold we have

$$\frac{(G(\hat{\theta}) - \theta)\sqrt{n(n-1)}}{p\sqrt{\sum_{i=1}^{n} [G_i(\hat{\theta}) - \hat{\theta})]^2}} \xrightarrow{\mathcal{L}} N(0,1) .$$

Proof: The fact that $G(\hat{\theta})$ L.O.B.E. $\hat{\theta}$ follows from Theorem I.3.3. Moreover in the proof of that theorem we saw that

$$\lim_{n\to\infty} \frac{R(n)}{1 - R(n)} \cdot \frac{1}{n} = \frac{1}{p} ,$$

and the rest of the theorem then follows by Theorem 2.4.

In the context of the preceeding asymptotic theorems we should mention another facet of the selection of $R(n)$. First note by the proof of Theorem 2.3 that when $G(\hat{\theta})$ is defined as in that theorem we can write

$$s_G^2 = \frac{1}{n-1} \sum_{i=1}^{n} (G_i(\hat{\theta}) - G(\hat{\theta}))^2$$

$$= \frac{R^2(n)}{(1 - R(n))^2} \frac{1}{(n - 1)} \sum_{i=1}^{n} \left[f(\bar{x}_n^i) - \frac{1}{n} \sum_{j=1}^{n} f(\bar{x}_{n-1}^j) \right]^2$$

$$= \left[\frac{R(n)}{(1 - R(n))(n - 1)} \right]^2 (n - 1) \sum_{i=1}^{n} \left[f(\bar{x}_n^i) - \frac{1}{n} \sum_{j=1}^{n} f(\bar{x}_{n-1}^j) \right]^2 . \qquad (2.13)$$

However if

$$G(\hat{\theta}) = J(\hat{\theta}) \quad ,$$

we see by (2.12) and (2.13) that

$$S_G^2 = (n - 1) \sum_{i=1}^{n} \left[f(\bar{x}_n^i) - \frac{1}{n} \sum_{j=1}^{n} f(\bar{x}_{n-1}^j) \right]^2 . \qquad (2.14)$$

Thus if we denote the right-hand side of equation (2.14) by S_J^2 we see that

$$S_G^2 = \left[\frac{R(n)}{(1 - R(n))(n - 1)} \right]^2 S_J^2 . \qquad (2.15)$$

From (2.15) and our previous comments it is clear that in Theorems 2.1 and 2.4 we can take $\alpha = 1$ if we replace S_G^2 by S_J^2 and hence we have the following theorem.

<u>THEOREM 2.6</u>. Let $\hat{\theta}$ be defined as in Theorem 2.4. Then

$$\frac{(G(\hat{\theta}) - \theta)\sqrt{n(n-1)}}{\sqrt{\sum_{i=1}^{n} [J_i(\hat{\theta}) - J(\hat{\theta})]^2}} \underset{\mathcal{L}}{\rightarrow} N(0,1) , \qquad (2.16)$$

as $n \to \infty$.

Theorem 2.6 suggests an interesting question. Since one can obtain the asymptotic normality property using S_G^2 or S_J^2 to estimate the variance of $G(\hat{\theta})$, which estimator is preferable? To consider this note that the approximate confidence interval of size $(1 - \beta)$ generated by (2.11) is

$$G(\hat{\theta}) - t_{\beta/2} \frac{S_G}{\alpha\sqrt{n}} \le \theta \le G(\hat{\theta}) + t_{\beta/2} \frac{S_G}{\alpha\sqrt{n}} ,$$

or

$$G(\hat{\theta}) - t_{\beta/2} \left| \frac{R(n)}{(1 - R(n))(n-1)\alpha} \right| \frac{S_J}{\sqrt{n}} \le \theta$$

$$\le G(\hat{\theta}) + t_{\beta/2} \left| \frac{R(n)}{(1 - R(n))(n-1)\alpha} \right| \frac{S_J}{\sqrt{n}} , \qquad (2.17)$$

where $t_{\beta/2}$ is defined by

$$\frac{1}{\sqrt{2\pi}} \int_{t_{\beta/2}}^{\infty} e^{(-1/2)x^2}\, dx = \beta/2 .$$

In practice we will use the corresponding point of the t distribution with (n - 1) degrees of freedom but this does not affect us here and we shall delay any further discussion of it until later. Now if L is the length of the confidence interval defined by equation (2.17) we see that

$$E[L] = \frac{2t_{\beta/2}}{\sqrt{n}} \left| \frac{R(n)}{(1 - R(n))(n - 1)\alpha} \right| E[S_J] . \tag{2.18}$$

However, since

$$\frac{R(n)}{(1 - R(n))(n - 1)} \equiv 1$$

when

$$R(n) = \frac{n - 1}{n}$$

we have the following theorem.

<u>THEOREM 2.7</u>. Let $\hat{\theta}$ be defined as in Theorem 2.4 and let L be the length of the confidence interval given by equation (2.17). Moreover let L_1 be the

length of the confidence interval defined by

$$G(\hat{\theta}) - t_{\beta/2} \frac{S_J}{\sqrt{n}} \leq \theta \leq G(\hat{\theta}) + t_{\beta/2} \frac{S_J}{\sqrt{n}}, \tag{2.19}$$

i.e.,

$$L_1 = \frac{2}{\sqrt{n}} t_{\beta/2} S_J . \tag{2.20}$$

Then

$$E[L] \leq E[L_1] \tag{2.21}$$

if and only if

$$\left| \frac{R(n)}{(1 - R(n))(n - 1)\alpha} \right| \leq 1. \tag{2.22}$$

Although Theorem 2.7 is interesting and suggests a possible answer to the question of how we should estimate the variance of $G(\hat{\theta})$, it does not completely settle the question. This is true because it is not clear that an approximate confidence interval should be considered preferable on the basis of a smaller expected length. However in those cases in which $G(\hat{\theta})$ is L.O.B.E. $\hat{\theta}$ one might consider the shorter confidence interval the more satisfactory one. Again this is certainly not an absolute and some empirical work would be interesting in light of the above discussion.

THEOREM 2.8. Let p be defined as in Theorem I.3.3 and

$$R(n) = \left(\frac{n - 1}{n}\right)^p .$$

Then $G(\hat{\theta})$ is L.O.B.E. $\hat{\theta}$ and furthermore

(i) if $p > 1$, $E[L] < E[L_1]$,

and

(ii) if $0 < p < 1$, $E[L] > E[L_1]$.

Proof: The first part of the theorem is part of the result of Theorem I.3.3. To establish the remaining part we must show that

$$\left|\frac{R(n)}{(1 - R(n))(n - 1)\alpha}\right| \begin{cases} < 1, & \text{if } p > 1 \\ = 1, & \text{if } p = 1 \\ > 1, & \text{if } 0 < p < 1 \end{cases} .$$

From the proof of Theorem 2.5, $\alpha = 1/p$ and since $R(n) < 1$ we have

$$\left|\frac{R(n)}{(1 - R(n))(n - 1)\alpha}\right| = \frac{p(n - 1)^{p-1}}{[n^p - (n - 1)^p]} . \qquad (2.23)$$

Now denote the right side of (2.23) by $g(p,n)$. Then it is clear that $g(1,n) \equiv 1$. Moreover

$$g(p,n) = 1$$

yields

$$p(n - 1)^{p-1} = n^p - (n - 1)^p$$

or

$$\frac{p + n - 1}{n} = \left(\frac{n}{n - 1}\right)^{p-1}$$

so that

$$\log (p + n - 1) = p \log \left(\frac{n}{n - 1}\right) + \log (n - 1) \quad . \qquad (2.24)$$

However, the only value of $p > 0$ for which (2.24) holds is $p = 1$. Hence $g(p,n) \neq 1$ if $p \neq 1$. But it is easy to show that $g(1/2,n) > 1$ and $g(2,n) < 1$. Since $g(p,n)$ is a continuous function of p, it therefore follows that

$$g(p,n) \begin{cases} < 1, & \text{if } p > 1 \\ > 1, & \text{if } 0 < p < 1 \ , \end{cases} \qquad (2.25)$$

and the theorem is established.

Before proceeding to the examples we should include one final theorem which is a mild generalization of a theorem due to Brillinger [7].

THEOREM 2.9. Let $X_1, X_2, \ldots, X_n$ be a random sample from a distribution with p.d.f. $f_X(\cdot;\theta)$, $\hat{\theta}^2$ be a maximum likelihood estimator of θ and $n = NM$. If N is held fixed and $\lim_{n\to\infty} R(n) \neq 1$ then the random variable

$$\frac{(G(\hat{\theta}) - \theta)\sqrt{N(N-1)}}{\sqrt{\sum_{i=1}^{N} [J_i(\hat{\theta}) - J(\hat{\theta})]^2}}$$

is asymptotically distributed as a Student's t with N - 1 degrees of freedom as $n \to \infty$.

Proof: By our assumptions, for values of $\hat{\theta}$ sufficiently close to θ we can write (here as in other places when convenient we use the same notation for the estimator and the estimate)

$$\sum_{i=1}^{n} \frac{\partial \log f_X(x_i,\theta)}{\partial\theta}\bigg|_{\theta=\hat{\theta}} = \sum_{i=1}^{n} \frac{\partial \log f_X(x_i,\theta)}{\partial\theta}$$

$$+ (\hat{\theta} - \theta) \sum_{i=1}^{n} \frac{\partial^2 \log f_X(x_i,\theta^*)}{\partial\theta^2} = 0, \qquad (2.26)$$

where θ^* lies between $\hat{\theta}$ and θ. By (2.26) and the consistency of $\hat{\theta}$ we can say

$$\sqrt{n}\,(\hat{\theta} - \theta) = \left[\sum_{i=1}^{n} \frac{\partial \log f_X(x_i,\theta)}{\partial\theta} \bigg/ \sqrt{n}\, I\right] + Y_n \;, \qquad (2.27)$$

where I is Fisher's information, i.e.,

$$I = -E\left[\frac{\partial^2 \log f_X(X,\theta)}{\partial\theta^2}\right],$$

and Y_n is a random variable which approaches zero in probability as $n \to \infty$.

By (2.27) we can also write

$$\sqrt{M(N-1)}\,(\hat{\theta}^i - \theta) = \left[\sum_{\substack{j=1\\ j\notin S_i}}^{n} \frac{\partial \log f_X(X_j,\theta)}{\partial\theta} \Big/ I\sqrt{M(N-1)}\right] + Y_{n,i}, \tag{2.28}$$

where $Y_{n,i} \underset{p}{\to} 0$ and S_i is the set of indices of the X's belonging to the ith group.

From (2.27) and (2.28) we have

$$G_i(\hat{\theta}) - \theta = \frac{1}{(1-R(n))I}\left[\frac{1}{n}\sum_{j=1}^{n}\frac{\partial \log f_X(X_j,\theta)}{\partial\theta} - \frac{R(n)}{M(N-1)}\sum_{\substack{j=1\\ j\notin S_i}}^{n}\frac{\partial \log f_X(X_j,\theta)}{\partial\theta}\right]$$

$$+ \frac{Y_n}{(1 - R(n))\sqrt{n}} - \frac{R(n)}{(1 - R(n))\sqrt{M(N - 1)}} Y_{n,i} .$$

Now noting that

$$\sum_{j=1}^{n} = \sum_{\substack{j=1 \\ j \notin S_i}}^{n} + \sum_{j \in S_i} ,$$

we can write

$$G_i(\hat{\theta}) - \theta = \frac{1}{(1 - R(n))I} \left(\frac{1}{NM} - \frac{R(n)}{M(N - 1)}\right)$$

$$\cdot \sum_{\substack{j=1 \\ j \notin S_i}}^{n} \frac{\partial \log f_X(X_j, \theta)}{\partial \theta}$$

$$+ \frac{1}{(1 - R(n))In} \sum_{j \in S_i} \frac{\partial \log f_X(X_j, \theta)}{\partial \theta}$$

$$+ \frac{Y_n}{(1 - R(n))\sqrt{n}} - \frac{R(n)}{(1 - R(n))\sqrt{M(N - 1)}} Y_{n,i} . \tag{2.29}$$

Thus averaging over i we obtain

$$G(\hat{\theta}) - \theta = \frac{1}{In} \sum_{j=1}^{n} \frac{\partial \log f_X(X_j, \theta)}{\partial \theta} + \frac{Y_n}{(1 - R(n))\sqrt{n}}$$

$$- \frac{R(n)\overline{Y}_n}{(1 - R(n))\sqrt{M(N - 1)}} , \qquad (2.30)$$

where

$$\overline{Y}_n = \frac{1}{N} \sum_{i=1}^{N} Y_{n,i} .$$

We may now see from (2.30) that if N is held fixed and $M \to \infty$, we are assured by the central limit theorem that the random variable $\sqrt{n}\,(G(\hat{\theta}) - \theta)$ is asymptotically $N(0,\sigma^2)$, where

$$\sigma^2 = \frac{1}{I} .$$

To complete the theorem we note that if $R(n) = (N - 1)/N$, the coefficient of the first term in (2.29) is zero and hence the pseudovalues of $J(\hat{\theta})$ tend to be independent $N(0,N/I)$ random variables as $n \to \infty$ and the remainder of the theorem follows.

3. Approximate Confidence Intervals and Tests

Occasionally estimation problems arise in a setting for which there are no sound reasons for any specific assumptions as to the particular family of distributions which govern the sample information. There are also situations where the nature of the underlying family of distributions may be assumed, but the calculated statistic is so complicated as to virtually eliminate the possibility of deriving its distribution in a form useful to any inference problem. These circumstances generate a requirement for robust procedures for constructing approximate confidence intervals. The theory given in Section 2 provides us with the justification for employing the J and G estimators and their pseudovalues to construct interval estimates relying on the Student t distribution as the appropriate approximation. It was mentioned in connection with the interval constructed in equation (2.17) that $t_{\beta/2}$ should be taken to be the $(1 - \beta/2)$ point of the t distribution with $(n - 1)$ degrees of freedom. The ramification of the conjectured approximate independence of the pseudovalues $G_i(\hat{\theta})$ is that it is reasonable to regard $G(\hat{\theta})$ and $[1/(n - 1)]\sum[G_i(\hat{\theta}) - G(\hat{\theta})]^2$ as a sample mean and sample variance. Hence, with a heavy reliance on the central limit theorem and the general robustness of the t statistic, we may work with

$$\frac{\alpha\sqrt{n}\,[G(\hat{\theta}) - \theta]}{\sqrt{\dfrac{1}{n-1}\displaystyle\sum_{i=1}^{n}[G_i(\hat{\theta}) - G(\hat{\theta})]^2}} \tag{3.1}$$

as if it were distributed as a Student t with $(n - 1)$ degrees of freedom. The asymptotic normality, which was demonstrated for the general setting in Theorem 2.4, merely serves to bolster our confidence in the approximation.

Before proceeding to some specific examples let us consider two modifications to the practice which we just discussed. If the nature of the estimator $\hat{\theta}$ is such that the pseudovalues, $G_i(\hat{\theta})$

do not assume n distinct values, then we should reduce the degrees of freedom for the t distribution accordingly. Consider the extreme example of this provided by Example 3.3 of Chapter I involving the largest order statistic for a sample of size n. In this example there are only two distinct values assumed by the n pseudovalues. The stability of

$$S_J^2 = \frac{1}{n - 1} \sum_{i=1}^{n} [J_i(X_{(n)}) - J(X_{(n)})]^2 \qquad (3.2)$$

is no greater than that of a squared difference between two independent observations. This should prompt us to use only a single degree of freedom for the t statistic which might be constructed using the S_J of this example. The general rule proposed by Mosteller and Tukey [32] is as follows. Count the number of different values which appear as pseudovalues, subtract one from this, and use the result as the adjusted degrees of freedom for the approximate t. This rule of thumb must be qualified to exclude adjustment for "legitimate ties." If carrying more decimal places in the recording process or in the computation would very likely have made two pseudovalues different, then they should not be considered "the same" in applying the rule.

The investigation of the second possible modification has not been pursued to any depth as of yet, but may merit further study. In studying the effect of intraclass correlation on the confidence coefficients of several common types of confidence intervals which were derived under the assumption of random samples, Walsh [47] demonstrates the deleterious effect of positive correlation. To correct the t statistic when each pair of variables in the sample have correlation ρ, the multiplying factor is

$$\sqrt{\frac{1-\rho}{1+(n-1)\rho}} \quad ,$$

i.e., when the n values $x_1, \ldots, x_n$ represent a single observation of a multivariate normal population for which each of the n variables has mean μ, variance σ^2, and the correlation between two variables is ρ, then

$$\frac{(\bar{X}-\mu)\sqrt{n(n-1)}}{\sqrt{\sum_{i=1}^{n}(X_i-\bar{X})^2}} \sqrt{\frac{1-\rho}{1+(n-1)\rho}}$$

has the Student t distribution with n - 1 degrees of freedom. If the approximate independence of the pseudovalues is more adequately modeled by taking $\rho = 1/n$ rather than $\rho = 0$, then we should multiply the quantity which we shall use for constructing approximate confidence intervals by the factor

$$\left(\frac{n-1}{2n-1}\right)^{1/2} .$$

This has the effect of increasing the length of the interval estimates. For example when $n = 11$, the confidence interval for μ, which one sets assuming that $\rho = 0$, should be increased by the multiple

$$\left(\frac{21}{10}\right)^{1/2} \cong 1.45 \quad , \tag{3.3}$$

if ρ is actually $1/n$.

To illustrate the error incurred by ignoring the fact that $\rho > 0$ it can be seen from a table given by Walsh that, if $n = 16$ and $\rho = 0.1$ and one sets a 95% confidence interval on μ assuming that $\rho = 0$, then the true confidence level is only 0.74. If the conjecture that $\rho = 1/n$ is more nearly correct than the assumption that $\rho = 0$, then the error incurred by ignoring ρ will not be this serious; but the dangers are evident. Depending upon the nature of the estimator and the underlying distribution, the true correlation of the pseudo-values will vary. Some sampling experiments should be conducted; but the use of $\rho = 1/n$ is probably conservative. Let us turn to a specific example.

Example 3.1. Suppose as in [32] that we have a random sample from an unknown distribution. The 11 values which were recorded are (0.1,0.1,0.1,0.4,0.5, 1.0,1.1,1.3,1.9,1.9,4.7). There is no reason to assume that the underlying distribution is normal and we wish to set confidence limits on the standard deviation σ. The point estimator which we decide to work with is the square root of the sample variance, i.e., we take

$$S = \sqrt{\frac{1}{n-1} \sum_{i=1}^{n} (X_i - \overline{X})^2} .$$

Since S^2 is an unbiased U statistic for σ^2 we shall invoke the consequences of Theorem 2.4 and use the pseudovalues, associated with jackknifing S, to set approximate confidence limits on σ. First we compute the sample standard deviation s = 1.34347. Next in the absence of any supplementary information we assume that the bias is $O(n^{-1})$ and take $R(n) = (n - 1)/n$. Table 3.1 gives some of the intermediate calculations leading to the jackknifed estimate, J(s), and the approximate confidence limits.

The jackknifed estimate is

$$J(s) = \frac{1}{11} \sum_{j=1}^{11} J_j(s) = 1.4894$$

and its estimated standard deviation

$$\sqrt{s_J^2} = \left[\frac{1}{10} \sum_{j=1}^{11} [J_j(s) - J(s)]^2 \right]^{1/2} = 0.6244 .$$

The two-sided confidence limits on σ are then calculated by

$$1.4894 \pm [t_{\beta/2}(10)](0.6244) ,$$

where $t_{\beta/2}(10)$ is the $(1 - \beta/2)$ point of the t distribution with 10 degrees of freedom. In particular we have

Confidence Level $(1 - \beta)$	Interval
2/3	(0.85, 2.13)
0.95	(0.10, 2.88)

TABLE 3.1

j	Observation x_j	s^j (x_j omitted)	Pseudovalue $J_j(s) = 11s - 10s^j$
1	0.1	1.36382	1.1400
2	0.1	1.36382	1.1400
3	0.1	1.36382	1.1400
4	0.4	1.38888	0.8894
5	0.5	1.39539	0.8243
6	1.0	1.41457	0.6325
7	1.1	1.41578	0.6204
8	1.3	1.41563	0.6219
9	1.9	1.39427	0.8355
10	1.9	1.39427	0.8355
11	4.7	0.70742	7.7040

The possibility exists that it would be advantageous to jackknife a function of the original estimator S, whose sampling distribution might be better suited to a jackknife procedure. For instance, log (S) is felt to be generally better behaved than S with respect to the tails and the symmetry of their respective sampling distributions. Other authors have proposed the use of variance stabilizing transformations prior to jackknifing (see Arvesen and Schmitz [4]). If the suggestion to jackknife log (S) is pursued for our example problem at hand, and the approximate confidence limits on log (σ) are converted to an interval for σ, we obtain

Confidence Level $(1 - \beta)$	Interval
2/3	(0.94, 3.29)
0.95	(0.44, 6.93)

Suppose that an investigator had decided to use an assumption of normality in this problem. We know that $(n - 1)S^2/\sigma^2$ is distributed as a χ^2 with $(n - 1)$ degrees of freedom. Hence one could use the 1/6 and 5/6 points of this distribution (which are 5.78 and 14.15, respectively) to set a 2/3 confidence interval on σ^2. These limits are easily computed using $s^2 = 1.805$ and the values just mentioned to yield

$$\frac{10(1.805)}{14.15} = 1.28 \quad \text{and} \quad \frac{10(1.805)}{5.78} = 3.12 \quad .$$

The corresponding interval for σ is (1.13, 1.77).

Alternatively, the range $w = 4.7 - 0.1 = 4.6$ might have been employed along with percentage points of the distribution of w/σ in normal samples of size 11. The interval for σ (with 2/3 confidence) obtained in this fashion is (1.17, 1.91). This result along with the three previous ones are displayed in Table 3.2.

TABLE 3.2

Source of Confidence Interval	2/3 Limits on σ
J(s)	(0.85, 2.13)
J(log s)	(0.94, 3.29)
$(n - 1)s^2/\sigma^2$	(1.13, 1.77)
w/σ	(1.17, 1.91)

Since it was known that this sample came from an exponential distribution with $\sigma = 1$, we might well wonder if the combination of a biased estimator and an erroneously short confidence interval could cause a serious departure of the true confidence level from the nominal confidence level. A simple sampling experiment along the lines of this example presented by Mosteller and Tukey yields some interesting comparisons of the various jackknife procedures with the approach under an invalid normality assumption.

Confidence intervals were generated from 1000 samples of size 11 from the exponential distribution with the single parameter equal to one. The observed percentage of the intervals which covered the value $\sigma = 1$ were recorded and are given as the estimated true confidence level. In all cases the nomial level is 0.90. The first three procedures given in Table 3.2 were used in the Monte Carlo experiment and three additional ones as well. Consideration of Theorem I.3.4 and a suspicion that $p < 1$ lead us to somewhat blindly select $m = 1/2$ for the order of the bias, i.e., the G estimator was employed with

$$R(n) = \left(\frac{n-1}{n}\right)^m = \left(\frac{10}{11}\right)^{1/2} = 0.95346 \quad ,$$

as opposed to $R = 10/11 = 0.90909$ for the J estimator. This leads to a value of $\alpha = 2$, as we saw in Theorem 2.5. Furthermore

$$\frac{R(n)}{1 - R(n)} \; \frac{1}{n-1} = 2.0488$$

and therefore the confidence interval centered on G(s) and using S_G^2 is only 1.0244 times as long as the one using S_J^2. The coverage of $G(s) \pm (t_{\beta/2}/\sqrt{n})S_J$ was recorded as well as that of $G(s) \pm (t_{\beta/2}/\alpha\sqrt{n})S_G$. This particular example seems to offer very little insight into those questions posed in connection with Theorems 2.7 and 2.8.

Finally the interval which resulted from modifying the approximate t statistic based on J(s) and S_J^2 to account for an assumed correlation

($\rho = 1/n$) between the pseudovalues was calculated. Recall that it was mentioned earlier in this section that such an interval (when $n = 11$) is 45% longer than the unmodified interval. The results of these Monte Carlo runs are given in Table 3.3. Since in each case we are estimating a binomial proportion from 1000 Bernouilli trials, the 90% confidence intervals about the estimates given are shorter than ±3%, i.e., the estimates in Table 3.3 are probably within 3% of true values.

TABLE 3.3

Source of Interval	Estimated True Confidence Level
$(n - 1)S^2/\sigma^2$	67.6
J(s)	72.0
J (log (s))	80.6
G(s)	73.6
$G(s) \pm (t/\sqrt{n})S_J$	73.1
J(s); $\rho = 1/n$	82.5

Example 3.2. In the previous example it is quite clear that the correlation between the pseudovalues in the jackknife can be the major source of error in the jackknife method for obtaining approximate confidence intervals. In fact this is precisely the reason for the breakdown in the jackknife method in an example given by Miller [30] on the preservation of normality. That is, in [30]

Miller showed that even though $(\hat{\theta}^1, \hat{\theta}^2, \ldots, \hat{\theta}^n)$ has a multivariate normal distribution with intraclass correlation ρ_0 and common variance σ_0^2, it does not follow that

$$\frac{\sqrt{n(n-1)}\quad [J(\hat{\theta}) - \theta]}{\left[\sum_{i=1}^{n} (J_i(\hat{\theta}) - J(\hat{\theta}))^2\right]^{1/2}} \tag{3.4}$$

is even approximately a t statistic. The source of this difficulty, as we just mentioned lies in the correlation of the pseudovalues. Since the pseudovalues are only approximately independent and since, as we have seen, even a small correlation can have a serious effect on our approximate confidence intervals, let us consider the problem further. For clarity we will consider Miller's example in a different, but essentially equivalent form.

Suppose that $(J_1(\hat{\theta}), J_2(\hat{\theta}), \ldots, J_n(\hat{\theta}))$ has a multivariate normal distribution with correlation matrix

$$\begin{bmatrix} 1 & \rho & \rho & \ldots & \rho \\ \rho & 1 & \rho & \ldots & \rho \\ \rho & \rho & \cdot & & \cdot \\ \vdots & \vdots & & \ddots & \vdots \\ \rho & \rho & \ldots & \rho & 1 \end{bmatrix} . \tag{3.5}$$

Then from the result of Walsh discussed earlier in this section, it follows that the statistic defined by (3.4) is distributed as ct, where t is a t statistic and

$$c = \sqrt{\frac{1 + (n - 1)\rho}{1 - \rho}} \quad . \qquad (3.6)$$

Examination of (3.6) clearly demonstrates that if ρ is large (i.e., close to 1), (3.4) will be radically different from a t statistic. Note that this effect is solely due to the correlation between the pseudovalues. That is, departure from the t distribution holds true for any sample of normal variates with correlation matrix (3.5) and is not peculiar to the jackknife structure. It follows then from this example and the previous one that the confidence level of our approximate confidence intervals may be suspect when high correlation is present in the pseudovalues. More study in this area would be interesting.

Example 3.3. Recall the discriminant function which was computed in Chapter II.4. The linear function of the observed rates of occurrence of the five words: "and"(x_1), "in"(x_2), "of"(x_3), "the"(x_4), and "to"(x_5) was originally calculated to be

$$D(\underline{x}) = -2.8367 - 0.0190x_1 + 0.0281x_2 + 0.05264x_3 - 0.0164x_4 + 0.0406x_5 \quad .$$

The jackknifed discriminant function (from Table

II. 4.4) is

$$D^*(\underline{x}) = -3.0141 - 0.0195x_1 + 0.0301x_2$$
$$+ 0.0547x_3 - 0.017x_4 + 0.0420x_5 \quad .$$

In the absence of normal theory the investigator would seldom have significance tests at hand for the coefficients of his linear discriminant function. In other words, some indication of the acceptability of the hypotheses that each of the coefficients are equal to zero is desired and without an assumption of multivariate normality of the vectors $\underline{X}_i$, an exact test will often not be available. However, using the pseudovalues of each coefficient which appear in the table of pseudodiscriminants and the asymptotic theory of this chapter we may construct a critical ratio for each coefficient, a_i^* of D*. Table 3.4 gives the jackknifed coefficients (a_i^*), the sample variances ($S^2_{a_i^*}$) of the five sets of pseudovalues (which were averaged to obtain the coefficients), then the standard error of the coefficient ($S_{a_i^*}/\sqrt{11}$) and the appropriate critical ratio. Note that the coefficient of x_3 (the frequency of occurrence of the word "of") is probably the only a_i^* which should be considered to be different from zero. This is proposed since $t_{0.95}(10) = 2.23$ and our five approximate tests are dependent but we shall not pursue the consequences of this. The purpose here is merely to exemplify the use of the technique discussed at length in

Chapter II. 4. to assess the variability of estimates rather than to estimate error rates.

TABLE 3.4

Coefficients a_i^*	-0.0195	0.0301	0.0547	-0.0167	0.0420
Variances $S^2_{a_i^*}(10)^2$	0.410	0.244	0.121	0.046	0.360
Standard error $S_{a_i^*}/\sqrt{n}$	0.0193	0.0149	0.0105	0.00645	0.0181
Critical ratio $\lvert a_i^* \rvert \sqrt{n}/S_{a_i^*}$	1.0	2.0	5.2	2.6	2.3

As we mentioned before in connection with this example, if D* and its associated approximate significance tests are used, then any estimates of misclassification probabilities should come from the U method applied to each pseudodiscriminant.

Example 3.4. Possibly the most well-known application of all the theory put forth here thus far is an alternative to the classical F test for equality of variances. When the underlying populations are normal the best tests on variances in the one and two sample problems are based on the χ^2 and F distributions. However, it has been known for some time that these tests are extremely sensitive to nonnormality. Miller [31] first presented a theoretical and empirical investigation of the approximate test (based on the pseudovalues of the ordinary jackknife) and its competitors, e.g., F, Box-Anderson,Levene, Box and Moses. The Monte Carlo studies given by Miller focus upon the small sample power functions for several tests of the hypothesis

$$H_0: \sigma_x^2 = \sigma_y^2$$

or equivalently

$$H_0: \sigma_y^2 / \sigma_x^2 = 1 \quad .$$

Because of the widely accepted claim that the log transformation applied to sums of squares can be beneficial, due to variance stabilization, and creation of a more normal looking distribution, i.e., reduction of the degree of asymmetry, the statistic which is jackknifed in this problem is $\log (S_y^2) - \log (S_x^2)$. The equivalent hypothesis under test with the jackknife statistic is simply

$$H_0: \log (\sigma_y^2 / \sigma_x^2) = 0 \quad .$$

The importance of symmetry of the distribution of the pseudovalues has not been stressed, but that symmetry does have an impact upon the goodness of the jackknife approximation to the t statistic, as we would expect. Arvensen and Schmitz [4] give some Monte Carlo comparisons of testing a variance ratio with and without the log transformation and the superiority of jackknifing the transformed statistic is well demonstrated there. We should note that the transformed statistic in the two sample problem does not fall under the assumptions underlying most of the theory of this chapter. Thus similar theorems concerning asymptotic distributions for the jackknife statistic produced from functions of several U statistics are needed here. Such results have been established by Arvesen and may be found in [3]. Additional work on testing ratios of scale parameters, which is related to that of Miller, appears in Shorack [45].

Consider now the following elementary variance components model:

$$Y_{ij} = \mu + a_i + e_{ij}, \quad i = 1,\ldots,I, \; j = 1,\ldots,M , \tag{3.7}$$

where μ is constant; $\{a_i\}$ are independent mean zero, variance σ_A^2 random variables; and the $\{e_{ij}\}$ are independent, mean zero, variance σ_e^2 random variables. The I + IM random variables $\{a_i\}$ and $\{e_{ij}\}$ are assumed to be mutually independent and to possess finite fourth moments.

We mentioned above that the F test of the hypothesis

$$H_0\colon \quad \eta = \sigma_A^2 / \sigma_e^2 \le \eta_0 \tag{3.8}$$

$$\text{vs.} \quad H_1\colon \quad \eta > \eta_0 \quad ,$$

is sensitive to the normality assumption on $\{a_i\}$ and $\{e_{ij}\}$. If the jackknife is to be a viable alternative it must do well when the data are normally distributed and be robust in the face of nonnormality.

Let MSA and MSE denote the between and within group mean squares, respectively. Next define the statistic

$$\hat{\theta} = \log \left(\frac{\text{MSA}}{\text{MSE}}\right) .$$

This may be used to test a hypothesis equivalent to (3.8), namely,

$$H_0: \quad \theta = \log (M\eta + 1) \leq \log (M\eta_0 + 1)$$

$$= \theta_0$$

$$\text{vs.} \quad H_1: \quad \theta > \theta_0 .$$

Now grouping the n = IM observations $\{Y_{ij}\}$ into the natural groups of size M, the method of Quenouille is applied by successively deleting the I groups of observations $\{(Y_{i1}, Y_{i2}, \ldots, Y_{im}),\ i = 1,2,\ldots,I\}$ in the computation of $\hat{\theta}$. The approximate t statistic derived from the jackknife with $R(n) = (n - 1)/n$ performs quite well here. Discussion of its use when the model of (3.7) is unbalanced may be found in [3] and a realistic example of the technique extended to a more complicated balanced design is given in [4].

A Monte Carlo study reported by Arvesen and Schmitz used the model

$$Y_{ij} = a_i + e_{ij}, \quad i = 1,\ldots,15, \quad j = 1,\ldots,3 \quad .$$

In other words, data were generated under (3.7) with $\mu = 0$, $I = 15$, and $M = 3$. Three distributions were employed for both the $\{a_i\}$ and $\{e_{ij}\}$, viz., normal, double exponential (kurtosis = 2), and uniform (kurtosis = -1.2). Table 3.5 gives the estimated power curves for the tests based on F and $J(\hat{\theta})$ for the hypothesis

$$H_0: \quad \eta = \sigma_A^2 \ / \ \sigma_e^2 \leq 1$$

$$\text{vs.} \quad H_1: \quad \eta > 1 \quad ,$$

at the $\alpha = 0.10$ significance level.

One thousand sets of $\{a_i\}$ and $\{e_{ij}\}$ were generated from each of the three distributions. They were obtained first with $\eta = 1$, and then scaled so that $\eta = 1.5, 2.5, 4, 6$. The jackknife was used in conjunction with the log transformation for each of the three distributions and also without the transformation when the data were normal. The resultingly poor power curve for J(MSA/MSE) emphasizes the desirability of reduced skewness through such transformations.

It is interesting to note how well the $J(\hat{\theta})$ test agrees with the F test on normal data.

TABLE 3.5

	$\eta = \sigma_A^2/\sigma_e^2$	1	1.5	2.5	4	6
Distribution						
Normal	F test (theoretical)	0.10	0.29	0.65	0.90	0.98
	F test (empirical)	0.098	0.269	0.655	0.900	0.979
	Jackknife ($J(\hat{\theta})$) (w/ transformation)	0.090	0.257	0.609	0.881	0.972
	Jackknife (J(MSA/MSE)) (w/o transformation)	0.022	0.080	0.263	0.469	0.700
Double Exponential	F test	0.183	0.334	0.567	0.764	0.887
	Jackknife (w/transformation)	0.096	0.218	0.420	0.616	0.767
Uniform	F test	0.085	0.253	0.690	0.951	0.994
	Jackknife (w/transformation)	0.093	0.308	0.733	0.947	0.993

VALUES OF THE MONTECARLO POWER FUNCTION FOR TESTING H_0 WITH $\alpha = 0.10$

Also, the nonrobustness of the F with respect to significance level is apparent in the case of double exponential distributions. It would appear that one might get significant results almost twice as often (under the null) as the nominal significance level states, i.e., Table 3.5 gives 0.183 as the observed significance level as opposed to 0.10. Hence the fact that the F test is more powerful at other values of η is accompanied by far too many Type I errors. There is little difference between the two tests for uniform data but $J(\hat{\theta})$ appears to be slightly better. In summary the jackknifed logarithm of the ratio of means squares behaves well with respect to significance level and power.

4. Finite Populations

When we are dealing with simple random sampling of infinite populations or finite populations with replacement we have argued that the assumption that $R(n) = (n - 1)/n$ is fairly robust. If however, we are sampling without replacement from a finite population of size N, then there may be no R which is so widely applicable since we are not dealing with a random sample. Let us consider a specific problem. Suppose that we wish to estimate the variance of the sample mean, $\bar{Y}$, based on a sample of size n. An obvious estimator to consider is

$$S_{\bar{Y}}^2 = \frac{1}{n(n-1)} \sum_{i=1}^{n} (Y_i - \bar{Y}_n)^2 ,$$

which is unbiased for var $(\bar{Y}_n)$ if the sample is i.i.d. Now it is well-known that for sampling without replacement from a finite population,

$$\text{var}\ (\bar{Y}_n) = (1 - \frac{n}{N})(\frac{1}{n})[\sum_1^N (Y_i - \bar{Y}_N)^2/(N-1)]$$
$$= (\frac{N-n}{N})\frac{\sigma^2}{n} \quad ,$$

and that

$$E\left[\frac{1}{n-1}\sum_1^n (Y_i - \bar{Y}_n)^2\right] = \frac{1}{N-1}\sum_1^N (Y_i - \bar{Y}_N)^2 \ ,$$

and hence

$$E[S_{\bar{Y}}^2] = \frac{\sigma^2}{n} = (\frac{N-n}{N})\frac{\sigma^2}{n} + \frac{\sigma^2}{N} \quad ,$$

where

$$\sigma^2 = \sum_{i=1}^N (Y_i - \bar{Y}_N)^2/(N-1) \quad .$$

Consequently the bias of this estimator does not depend upon the sample size. This is unusual but the circumstance of estimating a quantity which does depend upon n is also; so it behooves us to pursue this further. It may be seen from the details of Example I. 3.1 that here the second estimator obtained by the method of Quenouille is simply $S_{\bar{Y}}^2$ times $n/(n - 1)$. Therefore the bias of this second estimator is given by

$$E\left[\frac{n}{n-1} s_{\bar{Y}}^2\right] = \frac{\sigma^2}{n-1}$$

$$= \frac{N-n}{N}\frac{\sigma^2}{n}$$

$$+ \left[\frac{1}{n-1} - \frac{N-n}{N}\frac{1}{n}\right]\sigma^2 .$$

Consequently

$$R(n) = \frac{1/N}{[N + n(n-1)]/Nn(n-1)} = \frac{n(n-1)}{N + n(n-1)} .$$

Using this combining parameter, the generalized jackknife $G(s_{\bar{Y}}^2)$ is unbiased and given by

$$G(s_{\bar{Y}}^2) = [s_{\bar{Y}}^2 - \frac{n(n-1)}{N + n(n-1)}\frac{n}{n-1} s_{\bar{Y}}^2] / [1$$

$$- \frac{n(n-1)}{N + n(n-1)}]$$

$$= \frac{N + n(n-1) - n^2}{N + n(n-1) - n(n-1)} s_{\bar{Y}}^2$$

$$= \frac{N-n}{N} s_{\bar{Y}}^2 .$$

This is recognizable as the standard unbiased estimator for the variance of $\overline{Y}$ which obviously could have been obtained directly using the finite population correction.

This example brings up several interesting points. First, some caution must be exercised in obtaining the proper generalized jackknife when the estimated quantity is dependent upon the original sample size. Second, the bias of the original estimator need not depend upon n if a second estimator can be produced which will lead to an $R \neq 1$. Next, the estimator $S^2_{\overline{Y}}$ is not consistent, in the sense that it does not approach var $(\overline{Y})$ as $n \to N$. That is, if $n = N$, then $\overline{Y} = E(Y)$ and therefore var $(\overline{Y}) = 0$, but the value of the estimator is

$$\frac{1}{N(N-1)} \sum_{1}^{N} (Y_i - \overline{Y}_N)^2 \neq 0 \quad .$$

However, the finite population correction $(N - n)/N$ which arose here through the generalized jackknife has induced consistency in the above sense, i.e.

$$G(S^2_{\overline{Y}_N}) = 0 \quad ,$$

as it should be. For further comments concerning the use of jackknifing to produce consistency see [21]. Finally we are reminded that the quantity S^2_G introduced in Section 2 is employed to estimate the variance of a "sample mean," namely, the average of the pseudovalues. The results of

Section 2 are based on a conjecture that the pseudovalues, J_i, are approximately i.i.d. because of the strong resemblence which they bear to the observations, X_i. This being the case, the dependence among the sample values for a sample from a finite population without replacement should be viewed as being transmitted to the corresponding pseudovalues. In other words, the character of the "sample" of pseudovalues is approximately the same as the data. Consequently the finite population correction should be applied to S_G^2 whenever we are sampling a finite population of size N without replacement. This means that given any biased estimator $\hat{\phi}$ and its corresponding R we would produce an estimator $G(\hat{\phi})$ which has less bias and an estimate of its variance given by

$$\frac{N - n}{n} \frac{1}{n(n - 1)} \sum_{i=1}^{n} [G_i(\hat{\phi}) - G(\hat{\phi})]^2 .$$

For variations on this theme see Jones [22] and J. N. K. Rao [37].

CHAPTER IV
JACKKNIFING STOCHASTIC PROCESSES

1. Introduction

In the previous chapters we have considered a general approach to the problem of bias reduction for estimators based on independent identically distributed observations. In the process we found that the procedure yielded results which had more far reaching effects than one might have hoped for at the outset. The results to which we are referring are of course the asymptotic distribution properties of $G(\hat{\theta})$.

In this chapter we would like to extend the notions of our previous work concerning estimators of the form $f(\overline{X})$ to stochastic processes which retain the "flavor" of a random sample. We shall not however be so ambitious as to attempt to extend our methods to "white noise" but rather to stochastic processes which are at least piecewise continuous.

Before proceeding to specific definitions and theorems let us first consider once more the problem of Example 3.5 of Chapter I, that is, the problem of estimating the reliability function $r(\lambda;X) = e^{-\lambda x}$. In that example we assumed the number of failures in time T, N(T), formed a Poisson process. This assumption was made in order to investigate the relationship between $J(\hat{\theta})$ and $G(\hat{\theta})$ but was really not necessary, provided we had an alternative manner for selecting the parameter R(n). For the problem as we shall discuss it now, we again need not assume $\{N(t);\ t \in [0,\infty)\}$ is a Poisson process, but in order to be specific we shall continue to make that assumption. When we abandon the motivational nature of our discussion we will drop that requirement.

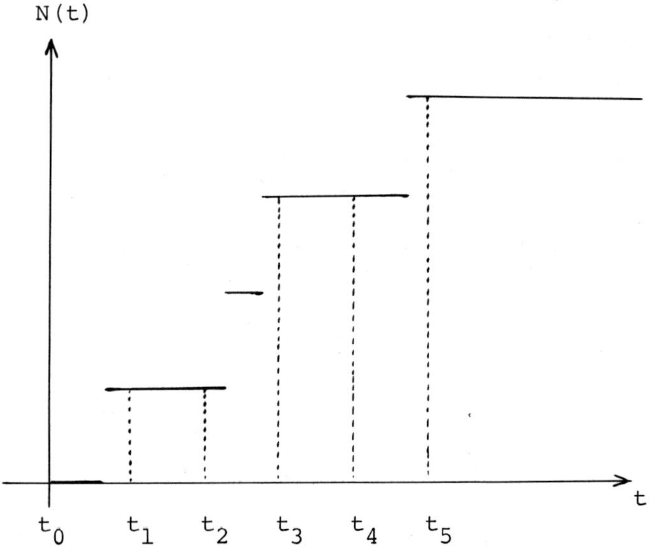

Figure 1.1

In Figure 1.1, the partitioning for n = 5 and $T = t_5$ of the process previously described is indicated. Thus

$$\hat{\lambda}(T) = \frac{N(T)}{T}$$

$$= \sum_{i=1}^{n} \frac{N(t_i) - N(t_{i-1})}{n\Delta t} ,$$

where $t_0 = 0$, and

$$r(\hat{\lambda};x) = \exp[-\hat{\lambda}(T)x]$$

$$= \exp\left[-\left(\sum_{i=1}^{n} \frac{N_i}{n\Delta t}\right) x\right] ,$$

where

$$N_i = N(t_i) - N(t_{i-1}) .$$

Moreover, still retaining the notation of Example 3.5,

$$r^i(\hat{\lambda};x) = \exp\left[-\left(\sum_{\substack{j=1 \\ j\neq i}}^{n} \frac{N_j}{n\Delta t}\right)x\right] ,$$

and since the N_i form a random sample we have

$$J[r(\hat{\lambda};x)] = nr(\hat{\lambda};x) - (n - 1)\ \overline{r^i(\hat{\lambda};x)} . \qquad (1.1)$$

Now in equation (1.1) it is clear that the value of J depends on n and hence different values of the partition size, Δt, of the time record give different values of J. This particular relationship was first investigated in [20] and later in [48]. In both cases it was found that for a fixed value of t, say t = T, the bias reduction properties of J were enhanced by increasing n. That is, the bias in $J[r(\hat{\lambda};x)]$, for $n\Delta t = T$, is a decreasing function of n. This immediately suggests considering the $\lim_{n\to\infty} J[r(\hat{\lambda};x)]$ with $n\Delta t = T$. This limit was obtained in [20,48] where it was shown that

$$\lim_{\substack{n\to\infty \\ n\Delta t=T}} J[r(\hat{\lambda};x)] = \exp[-\hat{\lambda}(T)x]\{1 - N(T)\exp[-\frac{X}{T}] + N(T) + \hat{\lambda}(T)x\} , \qquad (1.2)$$

and that the bias in the estimator defined by (1.2) is less than the bias in $J[r(\hat{\lambda};x)]$ for any finite n.

The above discussion suggests that, for certain types of stochastic processes, the notion of the jackknife has a natural extension which might be quite useful for bias reduction. This is precisely the case as was shown in [21] where the general notion of jackknifing stochastic processes was first introduced. In the next section we will begin a more thorough investigation of this concept. However, for a more complete study and the proofs of many of the theorems which we will presently give without proof, the reader is referred to [21] or [48].

2. The J(;n) Estimator

As was mentioned earlier we will restrict our discussion to stochastic processes which are at least piecewise continuous. In order to eliminate any confusion in our terminology we now state the following two definitions.

DEFINITION 2.1. A stochastic process $\{X(t) | t \in [a,b]\}$ will be said to be piecewise continuous (continuous) on [a,b] if, with probability one, every realization of $\{X(t) | t \in [a,b]\}$ is piecewise continuous (continuous).

DEFINITION 2.2. The process $\{X(t) | t \in [a,b]\}$ will be said to be of bounded variation on [a,b] if, with probability one, every realization of the process is of bounded variation on [a,b].

In addition to restricting the type of processes which we will consider it is also necessary to restrict the class of estimators. This is the purpose of the next definition.

DEFINITION 2.3. Let $\{X(t) | t \in S\}$ be a stochastic process defined over a set S containing the interval [a,b], and suppose that the probability law of X(t) depends on θ for every $t \in [a,b]$. Then for $t_1, t_2 \in [a,b]$ and $t_1 \neq t_2$, we define $\hat{\theta}(t_1,t_2)$ to be an estimator for θ of the form

$$\hat{\theta}(t_1,t_2) = \frac{I_X(t_2) - I_X(t_1)}{t_2 - t_1}, \qquad (2.1)$$

where $\{I_X(t)\,|\,t \in [a,b]\}$ is a piecewise continuous stochastic process which is completely determined by $\{X(t)\,|\,t \in [a,b]\}$.

Hereafter we will limit our discussion to estimators which are functions of $\hat{\theta}$, where $\hat{\theta}$ is defined by (2.1). Note that if $a = t_0 < t_1 < \ldots < t_n = b$ and $n(t_i - t_{i-1}) \equiv b - a$, then

$$\hat{\theta}(a,b) = \frac{1}{n} \sum_{i=1}^{n} \hat{\theta}(t_{i-1},t_i) \quad . \tag{2.2}$$

The following examples will be helpful in fixing the notion of the type of estimators we wish to consider.

Example 2.1. Let $\{N(t)\,|\,t \in [0,T]\}$ be the Poisson process discussed earlier. Then in the notation of Definition 2.3 we have $n(t_i - t_{i-1}) \equiv (b - a)$, $X(t) = N(t)$, $a = 0$, $b = T$, $\theta = \lambda$, and $I_X(t) = N(t)$. Therefore

$$\hat{\theta}(t_{i-1},t_i) = \hat{\lambda}(t_{i-1},t_i) = \frac{N(t_i) - N(t_{i-1})}{t_i - t_{i-1}}$$

$$= \frac{I_X(t_i) - I_X(t_{i-1})}{t_i - t_{i-1}} ,$$

$$\hat{\theta}(0,T) = \hat{\lambda}(0,T) = \frac{1}{n} \sum_{i=1}^{n} \hat{\lambda}(t_{i-1},t_i) = \frac{N(T)}{T} ,$$

and

$$f(\hat{\theta}) = r(\hat{\lambda};x) = \exp\,[-x\hat{\lambda}(0,T)] \quad .$$

Example 2.2. Let us examine Example 2.1 with $X(t) = N(t)/t$, $t \neq 0$, and $X(0) = 0$. Now consider the estimator $\hat{\lambda}(0,T)$ once more. Then, since

$$\hat{\lambda}(t_{i-1},t_i) = \frac{N(t_i) - N(t_{i-1})}{t_i - t_{i-1}} \ ,$$

in the notation of Definition 2.3 we would take

$$I_X(t) = N(t) = tX(t) \quad .$$

Example 2.3. Let

$$I_X(t) = \int_a^t X(y)\ dy \quad .$$

Then a familiar estimator for $E[X(t)]$ in a stationary ergodic process is

$$\hat{\theta}(a,b) = \frac{I_X(b) - I_X(a)}{b - a} = \frac{1}{b - a}\int_a^b X(y)\ dy \quad .$$

Thus if $n(t_i - t_{i-1}) \equiv b - a$ we have again

$$\hat{\theta}(a,b) = \frac{1}{n}\sum_{i=1}^{n} \frac{I_X(t_i) - I_X(t_{i-1})}{t_i - t_{i-1}} \quad .$$

We are now in a position to define the jackknife on functions of $\hat{\theta}$ for a regular partition of the interval [a,b], i.e., when $n(t_i - t_{i-1}) \equiv b - a$.

DEFINITION 2.4. Let $\{X(t) | t \in [a,b]\}$ be defined as in Definition 2.3. Then for any estimator $f(\hat{\theta})$ we define the J(;n) estimator of $f(\theta)$ by

$$J[f(\hat{\theta});n] = nf(\hat{\theta}) - \frac{n-1}{n} \sum_{i=1}^{n} f(\hat{\theta}_n^i) , \qquad (2.3)$$

where, for i = 1,2,...,n,

$$\hat{\theta} = \hat{\theta}(a,b)$$

$$b - a = n(t_i - t_{i-1})$$

$$\hat{\theta}_i = \hat{\theta}(t_{i-1},t_i)$$

and

$$\hat{\theta}_n^i = \frac{n}{n-1}\hat{\theta} - \frac{1}{n-1}\hat{\theta}_i . \qquad (2.4)$$

Clearly if one considers the $\hat{\theta}_i$'s a random sample of size n then J(;n) is the classical jackknife estimator which we have studied earlier. Thus for the process of Example 2.1 we have

$$J[r(\hat{\lambda};x);n] = nr(\hat{\lambda};x) - \frac{n-1}{n} \sum_{i=1}^{n} \exp\{-x[\frac{n}{n-1}\hat{\lambda} - \frac{1}{n-1}\hat{\lambda}_i]$$

$$= n\, r(\hat{\lambda};x) - \frac{n-1}{n} \exp\left[-\frac{nx}{n-1}\hat{\lambda}\right] \sum_{i=1}^{n} \exp\left[\frac{x}{n-1}\hat{\lambda}_i\right]$$

$$= r(\hat{\lambda};x)\left\{n - \frac{n-1}{n}\sum_{i=1}^{n}\exp\left[\frac{-x(\hat{\lambda}-\hat{\lambda}_i)}{n-1}\right]\right\} . \qquad (2.5)$$

From (2.5) it is clear that as $(b - a) \to \infty$, $J[r(\hat{\lambda};x);n]$ approaches $r(\lambda;x)$ in probability. Note that $J[r(\hat{\lambda};x);n]$ is the same estimator as $J[r(\hat{\lambda};x)]$ discussed in the introduction. We have however introduced the extra parameter at this point to emphasize that the integer n plays a somewhat different role in (2.3) than it does in jackknifing estimators based on random samples.

We shall now consider some properties of J(;n) by way of some obvious theorems.

THEOREM 2.1 Let $T = b - a$,

$$E[f(\hat{\theta})] = f(\theta) + B(T,\theta) \quad ,$$

and

$$E[f(\hat{\theta}^i_n)] = f(\theta) + B_{i,n}(T,\theta) \quad .$$

Then

$$E[J[f(\hat{\theta});n]] = f(\theta) + B(T,\theta)$$

$$+ (n-1)[B(T,\theta)$$

$$- B\left(\frac{n-1}{n}T,\theta\right)]$$

$$+ \frac{n-1}{n} \sum_{i=1}^{n} [B(\frac{n-1}{n} T,\theta) - B_{i,n}(T,\theta)] \; . \qquad (2.6)$$

Proof:

$$E[J(f(\hat{\theta});n)] = E[nf(\hat{\theta}) - \frac{n-1}{n} \sum_{i=1}^{n} f(\hat{\theta}_n^i)]$$

$$= f(\theta) + B(T,\theta)$$

$$+ \frac{n-1}{n} \sum_{i=1}^{n} [B(T,\theta) - B_{i,n}(T,\theta)] \; . \qquad (2.7)$$

Now by adding and subtracting B [(n - 1)/n]T,θ} in the summation of equation (2.7) the result follows.

COROLLARY 2.1. If in Theorem 2.1, $E[f(\hat{\theta}_n^i)] = E[f(\hat{\theta}_n^j)]$ for all i and j, then

$$E[J[f(\hat{\theta});n]] = f(\theta) + B(T,\theta) + (n-1)\Delta B(T,\theta) \; , \qquad (2.8)$$

where

$$\Delta B(T,\theta) = B(T,\theta) - B(T - \frac{T}{n},\theta) \; .$$

Proof: The proof follows from Theorem 2.1 by simply noting that $E[f(\hat{\theta}_n^n)] = f(\theta) + B\{[(n-1)/n]T,\theta\}$.

COROLLARY 2.2. If the process $\{I_X(t) \mid t \in [a,b]\}$ has stationary independent increments, then equation (2.8) holds.

From equation (2.8) we obtain the additional property established by the next theorem. In this theorem and hereafter we will always use the symbol T for the length of the interval [a,b], i.e., $T = b - a$.

THEOREM 2.2. If, $n > 1$ and $E[f(\hat{\theta}_n^i)] = E[f(\hat{\theta}_n^j)]$ for all i and j, then $J[f(\hat{\theta});n]$ is an unbiased estimator for $f(\theta)$ for all $n > 1$ if and only if $B(T,\theta) = D(\theta)/T$, for $n > 1$ where D is an arbitrary function of θ.

Proof: From (2.8), $J[f(\hat{\theta});n]$ is unbiased if and only if

$$B(T,\theta) + (n-1)\Delta B(T,\theta) = 0, \quad n > 1. \tag{2.9}$$

But equation (2.9) has solutions only of the form

$$B(T,\theta) = \frac{D(\theta)}{T},$$

and the theorem is established. Note that if $D(\theta) \equiv 0$, then $f(\hat{\theta})$ is unbiased and $J[f(\hat{\theta});n]$ is unbiased as well.

Similar results to Theorem I.3.3 can be obtained

for $J[f(\hat{\theta});n]$. In general however, the results are not quite as good. For example, suppose $B(T,\theta) = 0(T^{-2})$, i.e., there exists a function $C(\theta) \neq 0$ or $\pm\infty$ such that

$$\lim_{T\to\infty} T^2 \lim B(T,\theta) = C(\theta) .$$

Then when Corollary 2.1 holds

$$\frac{E[J(f(\hat{\theta});n)] - f(\theta)}{E[f(\hat{\theta})] - f(\theta)}$$

$$= \frac{B(T,\theta) + (n-1)\Delta B(T,\theta)}{B(T,\theta)}$$

$$= 1 + (n-1)\frac{B(T,\theta) - B\{[(n-1)/n]T,\theta\}}{B(T,\theta)}$$

$$= n - (n-1)\frac{B\{[(n-1)/n]T,\theta\}}{B(T,\theta)} .$$

But then

$$\left| \lim_{T\to\infty} \frac{E[J(f(\hat{\theta});n)] - f(\theta)}{E[f(\hat{\theta}) - f(\theta)]} \right|$$

$$= \left| n - (n-1) \lim_{T\to\infty} \frac{\{[(n-1)/n]T\}^2 B\{[(n-1)/n]T,\theta\}}{T^2 B(T,\theta)[(n-1)/n]^2} \right|$$

$$= 1 + \frac{1}{n-1}, \qquad (2.10)$$

and from (2.10) $f(\hat{\theta})$ is B.S.O.B.E. $J[f(\hat{\theta});n]$ for any finite n.

From the viewpoint of the order of bias and the notions surrounding that concept, which was discussed in Chapter I, (2.10) suggests that J(;n) may be increasingly effective as n increases. Moreover recall in our initial motivation, where $f(\hat{\theta}) = r(\hat{\lambda};x)$, that the bias in $J[f(\hat{\theta});n]$ is indeed a decreasing function of n. This, along with numerous other examples which could be given, suggests that the estimator which arises by considering the limit of $J[f(\hat{\theta});n]$ as $n \to \infty$, with T held fixed, may be of more general interest than J(;n). This turns out to be the case (at least at this stage of the development of the subject) and these observations lead to an estimator we shall refer to as the J_∞ estimator, which is the subject of the next section.

3. The J_∞ Estimator

Let us now give a formal definition of the J_∞ estimator, and consider its properties and the manner in which it relates to our discussion above.

DEFINITION 3.1. Let $\{X(t) | t \in S\}$ be a stochastic process defined over an index set S containing the interval [a,b]. Let $\{I_X(t) | t \in [a,b]\}$ be a piecewise continuous stochastic process which is completely determined by the process $\{X(t) | t \in [a,b]\}$. Let N_γ be the random variable defined by the number of discontinuities of I_X of size γ, observed on the interval [a,b]. Then for any real valued function f, differentiable over the range of $\hat{\theta}$, we define the J_∞ estimator by the following equation

$$J_\infty[f(\hat{\theta})] = f(\hat{\theta}) - \sum_{\gamma \in \Gamma} N_\gamma [f(\hat{\theta} - \tfrac{\gamma}{T}) - f(\hat{\theta}) + \tfrac{\gamma}{T} f'(\hat{\theta})] \tag{3.1}$$

where Γ is the set of all possible values of γ,

$$f'(\hat{\theta}) = \left. \frac{df(\theta)}{d\theta} \right|_{\theta=\hat{\theta}}$$

and by the notation $\sum_{\gamma \in \Gamma}$ we mean that for any realization x, the sum is to be taken over all γ for which N_γ has an observed value different from zero on [a,b]. Thus it is clear that the series in (3.1) is finite with probability 1. The following theorem establishes that $J_\infty[f(\hat{\theta})]$ is in fact the estimator obtained by proper consideration of the $\lim_{n \to \infty} J[f(\hat{\theta}); n]$. The proof is nontrivial and can be found in [21].

THEOREM 3.1. Let the stochastic process $\{I_X(t) | t \in [a,b]\}$ be as in Definition 3.1. Also let $\{I_X(t) | t \in [a,b]\}$ be of bounded variation on $[a,b]$, and for each $t \in [a,b]$, let the $I_X(t)$ be continuous at t with probability 1. Then

$$\lim_{n\to\infty} J[f(\hat{\theta});n] = J_\infty(f(\hat{\theta})) \quad . \tag{3.2}$$

Before proceeding to any additional properties of $J_\infty(f(\hat{\theta}))$, let us first give some further discussion of the random variable N_γ in Definition 3.1, and consider some examples.

Recall that

$$\hat{\theta}(t_{i-1},t_i) = \frac{I_X(t_i) - I_X(t_{i-1})}{t_i - t_{i-1}} \quad .$$

However the process $\{I_X(t) | t \in [a,b]\}$ is only assumed to be piecewise continuous and hence a particular realization, I_x, over $[a,b]$, I_x may have discontinuities. For instance, in the example where $I_X(t) = N(t)$, wherever a failure occurs I_x has a discontinuity of size 1, i.e., $\Gamma = \{1\}$ and $N_\gamma = N(T)$. From this discussion it is clear that a precise definition of N_γ is as follows. Let I_x be a realization of $\{I_X(t) | t \in [a,b]\}$ and let n_γ be the total number of points on $[a,b]$ such that

$$I_x(t^+) - I_x(t^-) = \gamma \neq 0 \quad ,$$

where t^+ and t^- indicate limits from the right and left, respectively. Then N_γ is the random variable whose values are the n_γ's. Some simplifications of (3.1) can now be given. We list them as examples.

Example 3.1. Suppose that, with probability one, every realization of $\{I_X(t) | t \in [a,b]\}$ is a step function. In this event

$$\sum_{\gamma\in\Gamma} N_\gamma \frac{\gamma}{T} = \hat{\theta}$$

and

$$J_\infty(f(\hat{\theta})) = f(\hat{\theta}) - \hat{\theta} f'(\hat{\theta}) - \sum_{\gamma\in\Gamma} N_\gamma [f(\hat{\theta} - \frac{\gamma}{T}) - f(\hat{\theta})] \quad . \qquad (3.3)$$

Example 3.2. Let $\Gamma = \{\gamma_0\}$. Then Γ has only one element and

$$J_\infty(f(\hat{\theta})) = f(\hat{\theta}) - N[f(\hat{\theta} - \frac{\gamma_0}{T}) - f(\hat{\theta}) + \frac{\gamma_0}{T} f'(\hat{\theta})] \qquad (3.4)$$

where N is defined by the total number of discontinuities of I_X observed on [a,b].

Example 3.3. Assume that with probability one, every realization of $\{I_X(t) | t \in [a,b]\}$ is a step function and $\Gamma = \{\gamma_0\}$. Then

$$J_\infty[f(\hat{\theta})] = f(\hat{\theta}) - \hat{\theta}f'(\hat{\theta}) - N[f(\hat{\theta} - \frac{\gamma_0}{T}) - f(\hat{\theta})] \tag{3.5}$$

where N is defined in Example 3.2.

Example 3.4. Assume that $\{I_X(t) | t \in [a,b]\}$ is continuous. Then

$$J_\infty(f(\hat{\theta})) = f(\hat{\theta}). \tag{3.6}$$

The result of equation (3.6) is expected since the notion we have extended, i.e., the classical jackknife, only takes advantage of the independent observations for the purpose of bias reduction. Thus when $I_X(t)$ is continuous, roughly speaking, none of the information for bias reduction on [a,b] is utilized. In some cases this is not bad since it may be the case that no useful information is available for that purpose. For example, in the Poisson case, when no failures are observed on [0,T] we get $J_\infty[r(\hat{\lambda};x)] = r(\hat{\lambda};x) = 1$ which is quite reasonable.

It was previously stated that in a general sense $J_\infty[f(\hat{\theta})]$ is superior to $J[f(\hat{\theta});n]$ with regard to its bias reduction properties. In the next few theorems we will establish sufficient properties of $J_\infty[f(\hat{\theta})]$ to clarify the meaning of such a remark.

<u>THEOREM 3.2</u>. Suppose for $T > T_0$

$$E[f(\hat{\theta})] = f(\hat{\theta}) + B(T,\theta)$$

and

$$E[f(\hat{\theta}_n^i)] = f(\theta) + B_{i,n}(T,\theta)$$

where $\hat{\theta}$ and $\hat{\theta}_n^i$ are defined in Definition 2.4. If $[\partial B(T,\theta)]/\partial T$ exists when $T > T_0$ and

$$\lim_{n\to\infty} E[J[f(\hat{\theta});n]] = E[J_\infty(f(\hat{\theta}))]$$

then

$$E[J_\infty(f(\hat{\theta}))] = f(\theta) + B(T,\theta) + T\,\frac{\partial B(T,\theta)}{\partial T} + \lim_{n\to\infty} \sum_{i=1}^{n} [B(\frac{n-1}{n}\,T,\theta) - B_{i,n}(T,\theta)]. \qquad (3.7)$$

Proof: From Theorem 2.1

$$E[J(f(\hat{\theta});n)] = f(\theta) + B(T,\theta) + (n - 1)[B(T,\theta) - B(\frac{n-1}{n}T, \theta)] + \frac{n-1}{n}\sum_{i=1}^{n}[B(\frac{n-1}{n}T,\theta) - B_{i,n}(T,\theta)] \quad . \tag{3.8}$$

But

$$\lim_{n\to\infty} (n - 1)[B(T,\theta) - B(\frac{n-1}{n}T,\theta)] = \lim_{n\to\infty} T(\frac{n-1}{n})[\frac{B(T,\theta) - B(T - (T/n),\theta)}{T/n}] = T\frac{\partial B(T,\theta)}{\partial T} \quad . \tag{3.9}$$

Using equations (3.8) and (3.9) and invoking the assumption that $\lim_{n\to\infty} E[J(f(\hat{\theta});n] = E[J_\infty(f(\hat{\theta}))]$, the result easily follows. We now state two obvious but useful corollaries.

<u>COROLLARY 3.1</u>. If Theorem 3.2 holds and in addition $E[f(\hat{\theta}_n^i)] = E[f(\hat{\theta}_n^j)]$ for all i and j, then

$$E[J_\infty(f(\hat{\theta}))] = f(\theta) + B(T,\theta) + T\frac{\partial B(T,\theta)}{\partial T} \quad . \tag{3.10}$$

<u>COROLLARY 3.2</u>. If Theorem 3.2 holds and $I_X(t)$ has stationary independent increments on [a,b], then equation (3.10) is satisfied.

One should note also that when Corollary 3.1 obtains if $f(\hat{\theta})$ is unbiased, $J_\infty[f(\hat{\theta})]$ is unbiased. A more general result which includes this observation is the following theorem.

<u>THEOREM 3.3</u>. Under the assumptions in Corollary 3.1, a necessary and sufficient condition for $J_\infty[f(\hat{\theta})]$ to be unbiased is that $B(T,\theta) = C(\theta)/T$, where $C(\theta)$ is an arbitrary function of θ.

Proof: The proof follows from equation (3.10) by noting that

$$B(T,\theta) + T \frac{\partial B(T,\theta)}{\partial T} = 0 \quad \text{for } T > T_0$$

if and only if $B(T,\theta) = C(\theta)/T$.

Generally the bias in $f(\hat{\theta})$ will not be of the form $C(\theta)/T$ and hence the J_∞ estimator will not eliminate all of the bias. For this reason we will now consider the behavior of $J_\infty[f(\hat{\theta})]$ when $B(T,\theta)$ differs from the form $C(\theta)/T$. In order to accomplish this let us make a slight modification of Definition I.3.1. In that definition we restricted the limit to be taken over the positive integers. At this time we would like to relax that requirement and allow the limit to be over the positive real numbers. With this convention, for example, $J_\infty[f(\hat{\theta})]$ B.S.O.B.E. $f(\hat{\theta})$ if

$$0 < \left| \lim_{T\to\infty} \frac{E[J_\infty(f(\hat{\theta})) - f(\theta)]}{E[f(\hat{\theta}) - f(\theta)]} \right| < 1 .$$

The adjustment of Definition I.3.1 which we speak of should now be obvious. We are therefore in a position to give an asymptotic characterization of the bias reduction properties of $J_\infty[f(\hat{\theta})]$ similar to those given for $J(\hat{\theta})$ in Chapter I. This is the purpose of the next theorem. We include the proof to aid us in understanding the implications of the theorem.

THEOREM 3.4. Let the conditions of Corollary 3.1 be satisfied and suppose that there exists a $p > 0$ such that

$$\lim_{T\to\infty} T^p B(T,\theta) = C(\theta) \neq 0 \text{ or } \pm \infty , \tag{3.11}$$

and $\lim_{T\to\infty} T^{p+1} [\partial B(T,\theta)]/\partial T$ exists. Then

(i) if $p = 1$, $J_\infty[f(\hat{\theta})]$ L.O.B.E. $f(\hat{\theta})$,

(ii) if $p < 2$, and $p \neq 1$, $J_\infty[f(\hat{\theta})]$ B.S.O.B.E. $f(\hat{\theta})$,

(iii) if $p = 2$, $J_\infty[f(\hat{\theta})]$ S.O.B.E. $f(\hat{\theta})$,

(iv) if $p > 2$, $f(\hat{\theta})$ B.S.O.B.E. $J_\infty[f(\hat{\theta})]$.

Proof: By Corollary 3.1,

$$\frac{E[J_\infty(f(\hat{\theta})) - f(\theta)]}{E[f(\hat{\theta}) - f(\theta)]} = \frac{B(T,\theta) + T[\partial B(T,\theta)/\partial T]}{B(T,\theta)}$$

$$= 1 + \frac{T[\partial B(T,\theta)/\partial T]}{B(T,\theta)} \quad . \qquad (3.12)$$

Now since $C(\theta) \neq 0$ and $p > 0$ it follows that $B(T,\theta) \to 0$ as $T \to \infty$. Thus since $\lim_{T\to\infty} T^{p+1}[\partial B(T,\theta)/\partial T]$ exists, we can apply L'Hospital's rule to obtain

$$C(\theta) = \lim_{T\to\infty} \frac{B(T,\theta)}{T^{-p}}$$

$$= \lim_{T\to\infty} \left(-\frac{1}{p}\, T^{p+1} \frac{\partial B(T,\theta)}{\partial T}\right) \quad . \qquad (3.13)$$

However, the second term in (3.12) can be rearranged as follows:

$$\frac{T[\partial B(T,\theta)/\partial T]}{B(T,\theta)} = \frac{T^{p+1}[\partial B(T,\theta)/\partial T]}{T^p B(T,\theta)} \quad .$$

Thus by (3.11) and (3.13) we have

$$\left| \lim_{T\to\infty} \frac{E[J_\infty(f(\hat{\theta})) - f(\theta)]}{E[f(\hat{\theta}) - f(\theta)]} \right| = \left| 1 + \frac{-pC(\theta)}{C(\theta)} \right|$$

$$= \left| 1 - p \right| . \qquad (3.14)$$

Applying the definition of L.O.B.E., B.S.O.B.E., etc. to (3.14) gives the desired results.

Before proceeding to some examples let us give a theorem which gives some justification to our earlier comments concerning $J_\infty[f(\hat{\theta})]$ and $J[f(\hat{\theta});n]$.

THEOREM 3.5. If Corollary 3.1 is satisfied and $J_\infty[f(\hat{\theta})]$ is an asymptotically unbiased estimator of $f(\theta)$ whose bias is a monotone function of T, then

$$\left| E[J_\infty(f(\hat{\theta})) - f(\theta)] \right| \le \left| E[J[f(\hat{\theta});n] - f(\theta)] \right|$$

for all n.

Proof: The proof follows by applying the mean value theorem and making a short argument. It can be found in [21].

We now consider some simple examples.

Example 3.5. Suppose again that $\{N(t);t \in [0,\infty)\}$ is a Poisson process such that $E[N(t)] = \lambda t$ and let $\hat{\lambda} = N(T)/T$. Now let $\hat{\lambda}^2$ be an estimator for λ^2. Then

$$E[\hat{\lambda}^2] = \lambda^2 + \frac{\lambda}{T} \quad .$$

Thus by Theorem 3.3, $J_\infty[\hat{\lambda}^2]$ is unbiased for λ^2 and from Example 3.3

$$J_\infty(\hat{\lambda}^2) = \hat{\lambda}^2 - 2\hat{\lambda}^2 - N(T)[(\hat{\lambda} - \frac{1}{T})^2 - \hat{\lambda}^2]$$

$$= \hat{\lambda}^2 - \frac{N(T)}{T^2} \quad . \qquad (3.15)$$

From (3.15) one can easily show $E[J_\infty(\hat{\lambda}^2)] = \lambda^2$ and hence establish directly, in this example, the validity of Theorem 3.3.

Example 3.6. Let $f(\hat{\theta})$ be an estimator for $f(\theta)$ such that

$$E[f(\hat{\theta})] = f(\theta) + \sum_{k=1}^{\infty} \frac{c_k}{(T - a_k)^{k+\delta}} \qquad (3.16)$$

where c_k and a_k are arbitrary functions of θ and δ is a constant such that $|\delta| < 1$. Consider the following questions. Under what conditions is (a) $J_\infty[f(\hat{\theta})]$ L.O.B.E. $f(\hat{\theta})$; (b) $J_\infty[f(\hat{\theta})]$ B.S.O.B.E. $f(\hat{\theta})$; (c) $J_\infty[f(\hat{\theta})]$ S.O.B.E. $f(\hat{\theta})$; and (d) $f(\hat{\theta})$ B.S.O.B.E. $J_\infty[f(\hat{\theta})]$? Solution: From (3.16) and Theorem 3.4, it is clear that $J_\infty[f(\hat{\theta})]$ L.O.B.E. $f(\hat{\theta})$ if and only if $c_1 \neq 0$ and $\delta = 0$. Now, let c_n be the first nonzero c_k in (3.16). Then by

Theorem 3.4, $J_\infty[f(\hat{\theta})]$ B.S.O.B.E. $f(\hat{\theta})$ if and only if $n + \delta < 2$. Similarly $J_\infty[f(\hat{\theta})]$ is S.O.B.E. $f(\hat{\theta})$ if and only if $n = 2$ and $\delta = 0$ or $c_k \equiv 0$. Finally $f(\hat{\theta})$ B.S.O.B.E. $J_\infty[f(\hat{\theta})]$ if and only if $n + \delta > 2$.

<u>Example 3.7</u>. We now return to the estimator

$$r(\hat{\lambda};x) = e^{-\frac{N(T)}{T}x}$$

discussed in the introduction. By Example (3.3), or our earlier discussion,

$$\begin{aligned} J_\infty[r(\hat{\lambda};x)] &= e^{-\hat{\lambda}x} + \hat{\lambda}xe^{-\hat{\lambda}x} - N(T)[e^{-(\hat{\lambda}-(1/T))x} \\ &\quad - e^{-\hat{\lambda}x}] \\ &= e^{-\hat{\lambda}x}\{1 - N(T)[e^{x/T} - 1 - \tfrac{x}{T}]\} . \end{aligned}$$

Now it is elementary to show that

$$E[r(\hat{\lambda};x)] = e^{-\lambda T(1 - e^{-x/T})} ,$$

and hence by Corollary 3.2

$$\begin{aligned} E[J_\infty(r(\hat{\lambda};x))] &= e^{-\lambda T(1-e^{(-x/T)})} + \\ &\quad + Te^{-\lambda T(1-e^{-x/T})}[-\lambda(1 - e^{-x/T}) \\ &\quad + \frac{\lambda x}{T} e^{-x/T}] \end{aligned}$$

$$= e^{-\lambda T(1-e^{-x/T})}[1 + \lambda T e^{-x/T}(1 + \frac{x}{T}) - \lambda T] .$$

Moreover

$$\lim_{T\to\infty} T\, B(T,\theta) = \lim_{T\to\infty} T\, (e^{-\lambda T(1-e^{-x/T})} - e^{-\lambda x})$$

$$= \lim_{T\to\infty} \frac{1}{T^{-1}} [e^{-\lambda x}[\exp(\lambda \sum_{n=2}^{\infty} (-1)^n \frac{x^n}{T^{n-1}\, n!})$$

$$-1]]$$

$$= \lim_{T\to\infty} \frac{1}{-T^{-2}} e^{-\lambda x}[\exp (\lambda \sum_{n=2}^{\infty} (-1)^n$$

$$\frac{x^n}{T^{n-1}\, n!})][\lambda \sum_{n=2}^{\infty} \frac{(-1)^{n+1}(n-1)x^n}{T^n\, n!}]$$

$$= \frac{1}{2} \lambda x^2 e^{-\lambda x} . \tag{3.17}$$

Thus from (3.17), p = 1 and by Theorem 3.4, $J_\infty[r(\hat{\lambda};x)]$ L.O.B.E. $r(\hat{\lambda};x)$.

Example 3.8. For the process of Example 3.3 let

$$f(\lambda) = \frac{1}{\lambda} .$$

Then

$$f(\hat{\lambda}) = \frac{1}{\hat{\lambda}}$$

and

$$J_\infty[f(\hat{\lambda})] = \frac{2T}{N(T)} - \frac{T}{N(T) - 1} .$$

Moreover

$$E[f(\hat{\lambda}) | N(T) > 0] = \sum_{n=1}^{\infty} \frac{T}{n} P[N(T) = n | N(T) > 0]$$

$$= \frac{Te^{-\lambda T}}{1 - e^{-\lambda T}} \sum_{n=1}^{\infty} \frac{(\lambda T)^n}{n(n!)}$$

$$= \frac{T}{e^{\lambda T} - 1} \int_0^{\lambda T} \frac{1}{x} (e^x - 1)dx ,$$

and

$$E[J_\infty(f(\hat{\lambda})) | N(T) > 1]$$

$$= \sum_{n=2}^{\infty} \left(\frac{2T}{n} - \frac{T}{n - 1} \right) P[N(T) = n | N(T) > 1]$$

$$= \frac{Te^{-\lambda T}}{1 - e^{-\lambda T} - \lambda Te^{-\lambda T}} \sum_{n=2}^{\infty} \left(\frac{2}{n} - \frac{1}{n - 1} \right) \frac{(\lambda T)^n}{n!}$$

$$= \frac{2T}{e^{\lambda T} - 1 - \lambda T} \int_0^{\lambda T} \frac{1}{x}(e^x - 1 - x)dx$$

$$- \frac{\lambda T^2}{e^{\lambda T} - 1 - \lambda T} \int_0^{\lambda T} \frac{1}{x^2} (e^x - 1 - x)dx \ .$$

Since $f(\hat{\lambda})$ and $J_\infty(f(\hat{\lambda}))$ are not defined on all of the range of $\hat{\lambda}$, our previous results do not apply. However,

$$\lim_{T\to\infty} T(E[f(\hat{\lambda})|N(T) > 0] - \frac{1}{\lambda}]$$

$$= \lim_{T\to\infty} \frac{\lambda T^2 \int_0^{\lambda T} \frac{1}{x}(e^x - 1)dx - T(e^{\lambda T} - 1)}{\lambda(e^{\lambda T} - 1)}$$

$$= \lim_{T\to\infty} \frac{2\lambda T \int_0^{\lambda T} \frac{1}{x}(e^x - 1)dx + \lambda T(e^{\lambda T} - 1) - (e^{\lambda T}-1)-\lambda Te^{\lambda T}}{\lambda^2 e^{\lambda T}}$$

$$= \lim_{T\to\infty} \frac{2\lambda \int_0^{\lambda T} \frac{1}{x}(e^x - 1)dx + 2\lambda(e^{\lambda T} - 1) - \lambda - \lambda e^{\lambda T}}{\lambda^3 e^{\lambda T}}$$

$$= \lim_{T\to\infty} \frac{2\lambda(e^{\lambda T} - 1) + \lambda^2 e^{\lambda T}}{\lambda^4 e^{\lambda T}}$$

$$= \frac{1}{\lambda^2} + \frac{2}{\lambda^3} \tag{3.18}$$

and

$$\lim_{T\to\infty} T^2 (E[J_\infty(f(\hat{\lambda}))|N(T) > 1] - \frac{1}{\lambda})$$

$$= \lim_{T\to\infty} \left[\frac{2\lambda T^3 \int_0^{\lambda T} \frac{1}{x}(e^x-1-x)\,dx - \lambda^2 T^4 \int_0^{\lambda T} \frac{1}{x^2}(e^x-1-x)\,dx}{e^{\lambda T} - 1 - \lambda T} \right.$$

$$\left. - \frac{T^2(e^{\lambda T} - 1 - \lambda T)}{e^{\lambda T} - 1 - \lambda T} \right]$$

$$= \lim_{T\to\infty} \left[\frac{6\lambda T^2 \int_0^{\lambda T} \frac{1}{x}(e^x-1-x)\,dx - 4\lambda^2 T^3 \int_0^{\lambda T} \frac{1}{x^2}(e^x-1-x)\,dx}{\lambda(e^{\lambda T} - 1)} \right.$$

$$\left. - \frac{2T(e^{\lambda T} - 1 - \lambda T) - \lambda^2 T^3}{\lambda^2 e^{\lambda T}} \right]$$

$$= \lim_{T\to\infty} \left[\frac{12\lambda T \int_0^{\lambda T} \frac{1}{x}(e^x-1-x)\,dx - 12\lambda^2 T^2 \int_0^{\lambda T} \frac{1}{x^2}(e^x-1-x)\,dx}{\lambda^2 e^{\lambda T}} \right.$$

$$\left. - \frac{2(e^{\lambda T} - 1 - \lambda T) - 5\lambda^2 T^2}{\lambda^2 e^{\lambda T}} \right]$$

$$= \lim_{T\to\infty} \left[\frac{12\lambda \int_0^{\lambda T} \frac{1}{x}(e^x-1-x)\,dx - 24\lambda^2 T \int_0^{\lambda T} \frac{1}{x^2}(e^x-1-x)\,dx}{\lambda^3 e^{\lambda T}} \right.$$

$$\left. - \frac{2\lambda(e^{\lambda T} - 1) - 10\lambda^2 T}{\lambda^3 e^{\lambda T}} \right]$$

$$= \lim_{T\to\infty} \frac{-24\lambda^2 \int_0^{\lambda T} \frac{1}{x^2}(e^x-1-x)\,dx - 12\frac{\lambda}{T}(e^{\lambda T}-1) - 2\lambda^2 e^{\lambda T} + 2\lambda^2}{\lambda^4 e^{\lambda T}}$$

$$= \lim_{T\to\infty} \frac{-24\frac{\lambda}{T^2}(e^{\lambda T}-1-\lambda T) - 12\frac{\lambda^2}{T} e^{\lambda T} + \frac{12\lambda}{T^2}(e^{\lambda T}-1) - 2\lambda^3 e^{\lambda T}}{\lambda^5 e^{\lambda T}}$$

$$= \frac{-2}{\lambda^2} . \tag{3.19}$$

Thus, from equations (3.18) and (3.19), $J(f(\hat{\lambda}))$ L.O.B.E. $f(\hat{\lambda})$, since those equations imply that the limit of the ratio of the biases of $J_\infty(f(\hat{\lambda}))$ and $f(\hat{\lambda})$ is zero.

4. Asymptotic Distribution Properties of $J_\infty[f(\hat{\theta})]$

In the previous section we have introduced the J_∞ estimator and shown it to have bias reduction properties analogous to those of the jackknife. In this section we will see that the same is true with respect to the asymptotic distributions properties of $J_\infty[f(\hat{\theta})]$. That is, when one considers the length of the sample record, T, as the sample size, one essentially obtains the same asymptotic distribution properties for $J_\infty[f(\hat{\theta})]$ as one does for the jackknife. In order to establish these asymptotic properties, however, we need an estimator for the variance of $J_\infty[f(\hat{\theta})]$. To accomplish this we will first establish an estimator for $\sigma^2 = \text{var}\ [\hat{\theta}(a,a + 1)]$ and use this result to obtain the appropriate estimator for var $[J_\infty(f(\hat{\theta}))]$. Before we proceed further let us make some comments regarding our notation. Since we will be concerned in this section with $\lim_{T\to\infty} \sqrt{T}[J_\infty(f(\hat{\theta})) - f(\theta)]$, where $T = b - a$ and a is fixed, it might be better to alter our notation for $J_\infty[f(\hat{\theta})]$ as follows.

$$J_\infty[f(\hat{\theta}(a,b))] = f(\hat{\theta}(a,b)) - \sum_{\gamma \varepsilon \Gamma} N_\gamma(T)[f(\hat{\theta}(a,b)) - \frac{\gamma}{T}) - f(\hat{\theta}(a,b)) + \frac{\gamma}{T} f'(\hat{\theta}(a,b))]. \tag{4.1}$$

However, this latter notation is rather awkward and therefore we will not use it except where it is necessary for clarity. That is, we will usually assume that our brief discussion here is sufficient warning to remind us that $\hat{\theta}$, and N_γ are functions of the record length T.

To obtain the desired estimator for σ^2 we return to the estimator $J[f(\hat{\theta});n]$ which led us to $J_\infty[f(\hat{\theta})]$. Recall that to obtain $J[f(\hat{\theta});n]$ we first subdivided the interval [a,b] into n subintervals of fixed length $\Delta T = T/n$.

Let us again consider a partition of the interval [a,b] but, for convenience in what follows, let us alter our method of partitioning as follows. Denote the greatest integer not exceeding nT, by [nT], define $t_i - t_{i-1} = 1/n$ when $i \leq [nT]$ (n fixed), and partition the interval as shown in Figure 4.1. Note that now n is the number of subintervals in the unit interval. Now suppose $\{I_X(t) | t \varepsilon [a,\infty)\}$ has stationary independent increments. Then $t_i = a + (i/n)$ and consequently for a given n, the $\hat{\theta}(a + ((i - 1)/n), a + (i/n))$ are independent identically distributed random variables when $i \leq [nT]$. Moreover $\text{var}\ [\hat{\theta}(a + ((i - 1)/n), a + (i/n))] = n\ \text{var}\ [\hat{\theta}(a, a + 1)]$ and

$$\hat{\theta}\left(a, a + \frac{[nT]}{n}\right) = \frac{1}{[nT]} \sum_{i=1}^{[nT]} \hat{\theta}\left(a + \frac{i-1}{n}, a + \frac{i}{n}\right).$$

Thus it is clear that for fixed n

$$\sigma^2_{n,T} = \frac{1}{n([nT] - 1)} \sum_{i=1}^{[nT]} \left[\hat{\theta}\left(a + \frac{i-1}{n}, a + \frac{i}{n}\right) - \hat{\theta}\left(a, a + \frac{[nT]}{n}\right)\right]^2 \qquad (4.2)$$

is consistent for σ^2, the variance of $\hat{\theta}(a, a + 1)$.

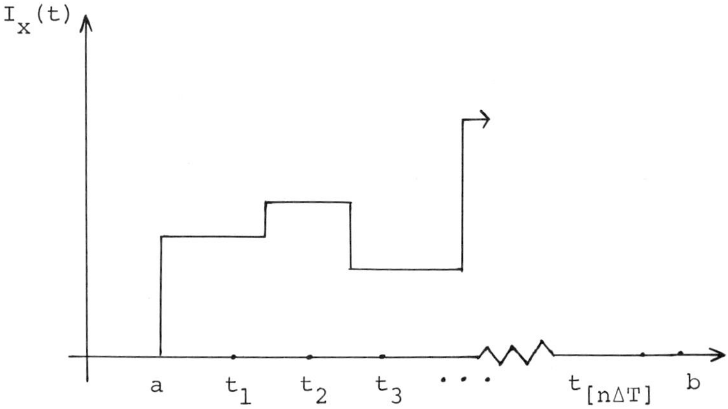

$$\Delta T = t_i - t_{i-1} = 1/n, \quad i \le [nT].$$

Figure 4.1

Since we obtained $J_\infty[f(\hat{\theta})]$ from the limit as $n \to \infty$ of $J[f(\hat{\theta});n]$ it would seem reasonable to consider $\lim_{n\to\infty} \hat{\sigma}^2_{n,T}$ in order to obtain an estimator for σ^2 independent of the partition size of the interval. In regard to this, it can be shown (see [48]) that

$$\lim_{n\to\infty} \hat{\sigma}^2_{n,T} = \frac{1}{T} \sum_{\gamma\varepsilon\Gamma} \gamma^2 N_\gamma \text{ , a.e. } . \tag{4.3}$$

We will denote the right side of (4.3) by $\hat{\sigma}^2_T$, i.e.,

$$\hat{\sigma}^2_T = \frac{1}{T} \sum_{\gamma\varepsilon\Gamma} \gamma^2 N_\gamma \quad .$$

Clearly by equation (4.2)

$$\lim_{n\to\infty} \lim_{T\to\infty} \hat{\sigma}^2_{n,T} = \lim_{n\to\infty} \sigma^2 = \sigma^2 \quad , \tag{4.4}$$

where the limits in (4.4) are limits in probability. Thus when these limits can be reversed, $\hat{\sigma}^2_T$ is consistent for σ^2. Rather than list conditions under which (4.4) holds we will simply assume $\hat{\sigma}^2_T$ is consistent for σ^2. Later we will give a result which often eliminates the need for this assumption. We now state a theorem which is the result for $J_\infty[f(\hat{\theta})]$ corresponding to Theorem III.2.2 for the jackknife. The proof can be found in [48].

THEOREM 4.1. Let $\{I_X(t) | t \in [a,\infty)\}$ be a stochastic process with stationary independent increments such that $E[\hat{\theta}(a,t)] = \theta$ for each $t \in (a,\infty)$ and suppose that f has a bounded second derivative in a neighborhood of θ. In addition suppose that $\hat{\sigma}_T^2 \underset{p}{\to} \sigma^2$ and that Γ is a bounded set. Then

$$\sqrt{T}\ [J_\infty(f(\hat{\theta}(a,a + T)) - f(\theta)]$$

$$\underset{\mathcal{L}}{\to} N(0,\sigma^2[f'(\theta)]^2) \quad , \qquad (4.5)$$

as $T \to \infty$, where $\mathcal{L}$ denotes convergence in law.

Although Theorem 4.1 is of some esthetic interest it is of no practical value without an estimator for the variance of $J_\infty[f(\hat{\theta}(a,a + T))]$. This can also be accomplished by subdividing the interval [a,b] and treating $J[f(\hat{\theta});n]$ as the ordinary jackknife. Thus following our previous development of the asymptotic distribution of the jackknife on a random sample, let

$$J_i[f(\hat{\theta});n] = n\ f(\hat{\theta}) - (n - 1)f(\hat{\theta}_n^i) \quad ,$$

and define

$$S_{n,T}^2 = \frac{1}{n(n - 1)} \sum_{i=1}^{n} \{J_i[f(\hat{\theta});n] - J[f(\hat{\theta});n]\}^2$$

as an estimator for the variance of $J[f(\hat{\theta});n]$. To obtain an estimator for the variance of $J_\infty[f(\hat{\theta})]$ we therefore consider the limit of $S_{n,T}^2$ as $n \to \infty$.

Under rather mild conditions this limit is given by (see [48]),

$$\lim_{n\to\infty} S^2_{n,T} = \sum_{\gamma\in\Gamma} N_\gamma [f(\hat{\theta} - \frac{\gamma}{T}) - f(\hat{\theta})]^2 \tag{4.6}$$

with probability one. If we then denote the right side of (4.6) by S^2_T one can further show [48] that when Theorem 4.1 holds

$$T\, S^2_T \underset{p}{\to} \sigma^2 [f'(\theta)]^2 \tag{4.7}$$

as $T \to \infty$. The upshot of our remarks and Theorem 4.1 is the following important result.

<u>THEOREM 4.2</u>. Under the conditions of Theorem 4.1,

$$\frac{J_\infty[f(\hat{\theta})] - f(\theta)}{\{\sum_{\gamma\in\Gamma} N_\gamma [f(\hat{\theta} - \frac{\gamma}{T}) - f(\hat{\theta})]^2\}^{1/2}} \underset{\mathcal{L}}{\to} N(0,1) \tag{4.8}$$

as $T \to \infty$.

Before exemplifying Theorem 4.2, let us make some remarks about our extension, $J_\infty[f(\hat{\theta})]$. It is quite clear that the manner in which we have extended the jackknife method to stochastic processes is somewhat restricted by the form of $\hat{\theta}$. However, at the present this is the extent to which this theory has been developed.

A rather natural question, which does not involve the form of $\hat{\theta}$ and which can be easily answered, is the following. Rather than extend the jackknife to stochastic processes, wouldn't it be better to extend the generalized jackknife? To answer this, note that the definition of $G[f(\hat{\theta});n]$ corresponding to $J[f(\hat{\theta});n]$ would be given by

$$G[f(\hat{\theta});n] = \frac{f(\hat{\theta}) - R(n)\,\overline{f(\hat{\theta}^i_n)}}{1 - R(n)}$$

$$= f(\hat{\theta}) + \frac{R(n)\ (n-1)[f(\hat{\theta}) - \overline{f(\hat{\theta}^i_n)}]}{[1 - R(n)](n-1)} \quad . \tag{4.9}$$

Consequently when the following limits exist in the appropriate sense we have

$$\lim_{n\to\infty} G[f(\hat{\theta});n] = f(\hat{\theta}) + \alpha \lim_{n\to\infty} \{(n-1)[f(\hat{\theta}) - \overline{f(\hat{\theta}^i_n)}]\} \quad ,$$

where, as in Chapter III,

$$\alpha = \lim_{n\to\infty} \frac{R(n)}{[1 - R(n)](n-1)} \quad .$$

However,

$$\lim_{n\to\infty} \{(n-1)[f(\hat{\theta}) - \overline{f(\hat{\theta}^i_n)}]\} = J_\infty[f(\hat{\theta})] - f(\hat{\theta}) \quad .$$

Hence

$$\lim_{n\to\infty} G[f(\hat{\theta});n] = f(\hat{\theta}) + \alpha\{J_\infty[f(\hat{\theta})] - f(\hat{\theta})\}$$

$$= (1-\alpha)f(\hat{\theta}) + \alpha J_\infty[f(\hat{\theta})]. \qquad (4.10)$$

Therefore, except for the fact that the estimators involved are not based on random samples, $G_\infty[f(\hat{\theta})]$ is a special case of $G(\hat{\theta}_1,\hat{\theta}_2)$ with

$$\hat{\theta}_1 = J\ [f(\hat{\theta})]$$

$$\hat{\theta}_2 = f(\hat{\theta})$$

and

$$R = \frac{\alpha - 1}{\alpha} \quad . \qquad (4.11)$$

Since it is usually difficult to obtain R through equation (4.11), let us now consider a different approach to selecting R. That is, let R be the ratio of the biases of $J_\infty[f(\hat{\theta})]$ and $f(\hat{\theta})$. Then when Corollary 3.1 holds, a reasonable selection for R would be obtained by letting

$$R = \frac{E[J_\infty(f(\hat{\theta})) - f(\theta)]}{E[f(\hat{\theta}) - f(\theta)]} = \frac{B(T,\theta) + T\frac{\partial}{\partial T}(B(T,\theta))}{B(T,\theta)} \, . \qquad (4.12)$$

But if $B(T,\theta) = h(T)C(\theta)$, then the right side of equation (4.12) becomes $[T\,h(T)]'/h(T)$ where the prime denotes differentiation with respect to T. Thus we would obtain

$$G_\infty[f(\hat{\theta})] = \frac{J_\infty[f(\hat{\theta})] - \dfrac{[Th(T)]'}{h(T)} f(\hat{\theta})}{1 - \dfrac{[Th(T)]'}{h(T)}}$$

$$= \frac{[Th(T)]'f(\hat{\theta}) - h(T)J_\infty[f(\hat{\theta})]}{Th'(T)}$$

$$= \frac{\begin{vmatrix} f(\hat{\theta}) & J_\infty[f(\hat{\theta})] \\ h(T) & [Th(T)]' \end{vmatrix}}{Th'(T)} .$$

The discussion above has essentially led us to an estimator which we obviously should refer to as the G_∞ estimator. This is the subject of the next section.

5. The G_∞ Estimator

DEFINITION 5.1. Let $h(T)$ be any differentiable function of T and suppose $Th'(T) \neq 0$. Then we define the G_∞ estimator by

$$G_\infty[f(\hat{\theta})] = \frac{\begin{vmatrix} f(\hat{\theta}) & J_\infty[f(\hat{\theta})] \\ h(T) & [Th(T)]' \end{vmatrix}}{Th'(T)}, \tag{5.1}$$

where

$$[Th(T)]' = \frac{d}{dT}[Th(T)] \tag{5.2}$$

and

$$h'(T) = \frac{dh(T)}{dT} . \tag{5.3}$$

It should be noted that if $h(T) = T^{-1}$, then $G_\infty[f(\hat{\theta})] = J_\infty[f(\hat{\theta})]$. Moreover if $B(T,\theta) = C(\theta)h(T)$, then $G_\infty[f(\hat{\theta})]$ is an unbiased estimator for $f(\theta)$. The establishment of $G_\infty[f(\hat{\theta})]$ makes it essentially always possible to produce an estimator which is L.O.B.E. $f(\hat{\theta})$ when the order of the bias in $f(\hat{\theta})$ is known and (4.12) holds. That is, when $B(T,\theta) = 0(T^{-p})$, taking $h(T) = T^{-p}$ in $G_\infty[f(\hat{\theta})]$ yields $G_\infty[f(\hat{\theta})]$ L.O.B.E. $f(\hat{\theta})$. This result is established by the next Theorem.

<u>THEOREM 5.1</u>. Assume Corollary 3.1 holds, and there exists a $p > 0$ such that

$$\lim_{T\to\infty} T^p B(T,\theta) = C(\theta) \neq 0, \pm \infty, \tag{5.4}$$

and

$$\lim_{T\to\infty} T^{p+1}[\partial B(T,\theta)/\partial T] \quad \text{exists} .$$

Then, for $h(T) = T^{-p}$, $G_\infty[f(\hat{\theta})]$ L.O.B.E. $f(\hat{\theta})$.

Proof: When $h(T) = T^{-p}$,

$$G_\infty[f(\hat{\theta})] = \frac{J_\infty[f(\hat{\theta})] + (p - 1)f(\hat{\theta})}{p}$$

and

$$E[G_\infty(f(\hat{\theta}))] = f(\theta) + B(T,\theta)$$

$$+ \frac{T}{p} \frac{\partial B(T,\theta)}{\partial T} . \tag{5.5}$$

This implies that

$$\frac{E[G_\infty(f(\hat{\theta})) - f(\theta)]}{E[f(\hat{\theta}) - f(\theta)]} = 1 + \frac{T}{p} \frac{\partial B(T,\theta)/\partial T}{B(T,\theta)} . \tag{5.6}$$

Now since $p > 0$ in (5.4), $B(T,\theta) \to 0$ as $T \to \infty$ and since also $\lim_{T\to\infty} T^{p+1} [\partial B(T,\theta)/\partial T]$ exists we can apply L'Hospital's rule to obtain

$$C(\theta) = \lim_{T\to\infty} \frac{B(T,\theta)}{T^{-p}} = \lim_{T\to\infty} \left(- \frac{\partial B(T,\theta)/\partial T}{p\ T^{-p-1}} \right). \tag{5.7}$$

Now, (5.7) may be used to evaluate the limit of (5.6) since

$$\lim_{T\to\infty} \left(\frac{T}{p} \frac{\partial B(T,\theta)/\partial T}{B(T,\theta)} \right) = \lim_{T\to\infty} \frac{[\partial B(T,\theta)/\partial T]/p\ T^{-p-1}}{B(T,\theta)/T^{-p}}$$

$$= \frac{-C(\theta)}{C(\theta)}$$

$$= -1 \quad .$$

Hence, we obtain

$$\lim_{T\to\infty} \frac{E[G_\infty(f(\hat{\theta})) - f(\theta)]}{E[f(\hat{\theta}) - f(\theta)]} = 0 \ ,$$

and the theorem is established.

Before giving an example we obtain a final result which shows that the asymptotic distribution of

$\sqrt{T}\ [G_\infty(f(\hat{\theta})) - f(\theta)]$ is essentially independent of $h(T)$.

<u>THEOREM 5.2</u>. Suppose the limit as $T \to \infty$ of $[Th(T)]'/h(T)$ exists and denote this limit by r. Then if $r \neq 1$ and Theorem 4.1 holds,

$$\sqrt{T}\ [G_\infty(f(\hat{\theta})) - f(\theta)] \underset{\mathcal{L}}{\to} N(0, \sigma^2(f'(\theta))^2) \qquad (5.8)$$

as $T \to \infty$.

Proof: Let

$$r(T) = \frac{[Th(T)]'}{h(T)} \quad .$$

Then by définition

$$\sqrt{T}\ \{G_\infty[f(\hat{\theta})] - f(\theta)\} = \sqrt{T}\left\{\frac{J_\infty[f(\hat{\theta})] - r(T)f(\hat{\theta})}{1 - r(T)} - f(\theta)\right\}$$

$$= \sqrt{T}\left\{\frac{J_\infty[f(\hat{\theta})] - f(\theta)}{1 - r(T)} + \frac{-r(T)[f(\hat{\theta}) - f(\theta)]}{1 - r(T)}\right\} .$$

But (see [48]) it can be shown that

$$\sqrt{T}\ [f(\hat{\theta}) - f(\theta)] \underset{\mathcal{L}}{\rightarrow} N(0,\sigma^2(f'(\theta))^2)$$

as $T \rightarrow \infty$. Also by Theorem 4.1 we know that

$$\sqrt{T}\ \{J_\infty[f(\hat{\theta})] - f(\theta)\} \underset{\mathcal{L}}{\rightarrow} N(0,\sigma^2(f'(\theta))^2)\ .$$

Moreover,

$$\begin{aligned} \operatorname{var} G_\infty[f(\hat{\theta})] = \frac{1}{[1 - r(T)]^2} \{ & \operatorname{var} J_\infty[f(\hat{\theta})] \\ & + (r(T))^2 \operatorname{var} f(\hat{\theta}) \\ & - 2r(T) \operatorname{cov} (J_\infty[f(\hat{\theta})], f(\hat{\theta}))\}\ , \end{aligned}$$

and therefore

$$\begin{aligned} \lim_{T\rightarrow\infty} \operatorname{var} G_\infty[f(\hat{\theta})] &= \frac{(1 + r^2 - 2r)\sigma^2(f'(\theta))^2}{(1 - r)^2} \\ &= \sigma^2(f'(\theta))^2\ . \end{aligned}$$

Combining these observations yields the desired result.

To this point, we have not given a specific example to demonstrate the possible uses of such theory as the normality theorems thus far developed. In addition we have not given the user any alternative to the assumption $\sigma_T^2 \underset{p}{\rightarrow} \sigma^2$, which has appeared in all of our results. The elimination of these objections is the purpose of the next section. The section is contributed by Professor T. A. Watkins.

6. Applications of the J_∞ Estimator (T. A. Watkins)

The result given in Theorem 4.2 relies heavily on the fact that $\hat{\sigma}_T^2$ is a consistent estimator of σ^2, and an analytical proof of this fact may be difficult in a great many situations. The following results prove useful in determining the consistency of $\hat{\sigma}_T^2$ in a variety of instances.

Lemma 6.1. Let $\{I_X(t) | t \in [a,\infty)\}$ be a piecewise continuous stochastic process, and for $t \geq a$ let $H_X(t) = I_X(t^+) - I_X(t^-)$. Suppose that Γ is countable and, for $t_1, t_2 \geq a$ with $t_1 \neq t_2$, $H_X(t_1)$ and $H_X(t_2)$ are independent identically distributed random variables such that

$$P[H_X(t) = \gamma | H_X(t) \neq 0] = p_\gamma \tag{6.1}$$

for each $\gamma \in \Gamma$. Also suppose that $\hat{P}_{\gamma,T}$ is a random variable defined by

$$\hat{P}_{\gamma,T} = \begin{cases} 0 & , \text{ when } N(T) = 0 \\ \dfrac{N_\gamma(T)}{N(T)} & , \text{ when } N(T) \neq 0 , \end{cases} \tag{6.2}$$

where $N(T) = \sum_{\gamma \varepsilon \Gamma} N_\gamma(T)$. If for any $k > 0$

$$\lim_{T \to \infty} P[N(T) > k] = 1 \tag{6.3}$$

then $\hat{P}_{\gamma,T}$ converges in mean square to p_γ as $T \to \infty$.

Proof: For each $\gamma \varepsilon \Gamma$, let $\{Y_\gamma(t) | t \varepsilon [a,\infty)\}$ be the stochastic process defined by

$$Y_\gamma(t) = \begin{cases} 1 , & \text{when } H_X(t) = \gamma \\ 0 , & \text{when } H_X(t) \neq \gamma . \end{cases} \tag{6.4}$$

Thus, applying (6.4), we see that

$$P[Y_\gamma(t) = 1 | H_X(t) \neq 0] = p_\gamma$$

and

$$P[Y_\gamma(t) = 0 \mid H_X(t) \neq 0] = 1 - p_\gamma .$$

Also, we have

$$N_\gamma(T) = \sum_{t\varepsilon[a,a+T]} Y_\gamma(t) \quad .$$

Thus

$$E[\hat{P}_{\gamma,T}] = E[E[\hat{P}_{\gamma,T}|N(T)]]$$

$$= \sum_{k=1}^{\infty} E\left[\frac{1}{N(T)} \sum_{t\varepsilon[a,a+T]} Y_\gamma(t)\,|\,N(T) = k\right]$$

$$\cdot P[N(T) = k] \quad .$$

Now we note that the conditional distribution of $\sum_{t\varepsilon[a,a+T]} Y_\gamma(t)$, given $N(T) = k$, is the binomial distribution with parameters p_γ and k, and hence,

$$E[\hat{P}_{\gamma,T}] = \sum_{k=1}^{\infty} p_\gamma P[N(T) = k]$$

$$= p_\gamma P[N(T) > 0] \quad . \tag{6.5}$$

Similarly,

$$E[(\hat{P}_{\gamma,T} - p_\gamma)^2] = \sum_{k=1}^{\infty} E[(\hat{P}_{\gamma,T} - p_\gamma)^2|N(T) = k]P[N(T)=k$$

$$= \sum_{k=1}^{\infty} \frac{p_\gamma(1 - p_\gamma)}{k} P[N(T) = k] \quad . \qquad (6.6)$$

Now, let $\varepsilon > 0$, and k_0 be a positive integer with $1/(k_0 + 1) < \varepsilon$. Then from equation (6.6), we have

$$E[(\hat{P}_{\gamma,T} - p_\gamma)^2] = \sum_{k=1}^{k_0} \frac{p_\gamma(1 - p_\gamma)}{k} P[N(T) = k]$$

$$+ \sum_{k=k_0+1}^{\infty} \frac{p_\gamma(1 - p_\gamma)}{k} P[N(T) = k]$$

$$\leq p_\gamma(1 - p_\gamma)P[1 \leq N(T) \leq k_0]$$

$$+ \varepsilon\, p_\gamma(1 - p_\gamma)P[N(T) > k_0] \quad . \qquad (6.7)$$

Thus, from (6.3), (6.5), and (6.7) we see that

$$\lim_{T\to\infty} E[\hat{P}_{\gamma,T}] = p_\gamma$$

and

$$\lim_{T\to\infty} E[(\hat{P}_{\gamma,T} - p_\gamma)^2] = 0 \ ,$$

which was to be shown.

<u>THEOREM 6.1</u>. Let the hypotheses of Lemma 6.1 be

satisfied, and let $\{I_X(t) | t \varepsilon [a,\infty)\}$ be a stochastic process such that, with probability one, every realization of $\{I_X(t) | t \varepsilon [a,\infty)\}$ is a step function and $\hat{\theta}(a,a + T) \to \theta$ as $T \to \infty$. If Γ is a finite set, and $H_X(t)$, given $H_X(t) \neq 0$, has a finite second moment and a nonzero mean, then

$$\hat{\sigma}_T^2 \underset{p}{\to} \frac{\theta \, E[(H_X(t))^2 | H_X(t) \neq 0]}{E[H_X(t) | H_X(t) \neq 0]} \tag{6.8}$$

as $T \to \infty$.

Proof: Since $N(T) \underset{p}{\to} \infty$, as $T \to \infty$, given $\varepsilon > 0$ there exists $T_\varepsilon > 0$ such that

$$P[N(T) > 0] > 1 - \varepsilon$$

whenever $T > T_\varepsilon$. Then if $T > T_\varepsilon$

$$\frac{1}{T} \sum_{\gamma \varepsilon \Gamma} \gamma^2 N_\gamma(T) = [\sum_{\gamma \varepsilon \Gamma} \gamma N_\gamma(T)/T]$$
$$\cdot \left[\sum_{\gamma \varepsilon \Gamma} \gamma^2 \frac{N_\gamma(T)}{N(T)} \Bigg/ \sum_{\gamma \varepsilon \Gamma} \gamma \frac{N_\gamma(T)}{N(T)} \right]$$

with probability greater than $1 - \varepsilon$. Now, since, with probability one, every realization of $\{I_X(t) | t \varepsilon [a,\infty)\}$ is a step function,

$$\frac{1}{T} \sum_{\gamma \varepsilon \Gamma} \gamma N_\gamma(T) = \hat{\theta}(a,a+T)$$

with probability one. Thus, when $T > T_\varepsilon$

$$\hat{\sigma}_T^2 = \left[\hat{\theta}(a,a+T) \sum_{\gamma \varepsilon \Gamma} \gamma^2 \frac{N_\gamma(T)}{N(T)} \Big/ \sum_{\gamma \varepsilon \Gamma} \gamma \frac{N_\gamma(T)}{N(T)} \right]$$

with probability greater than $1 - \varepsilon$. Now, since $\hat{\theta}(a,a+T) \underset{p}{\to} \theta$ and by Lemma 6.1, $[N_\gamma(T)/N(T)] \underset{p}{\to} p_\gamma$ as $T \to \infty$, and since Γ is finite, it is easily seen that

$$\hat{\sigma}_T^2 \underset{p}{\to} \frac{\theta \sum_{\gamma \varepsilon \Gamma} \gamma^2 p_\gamma}{\sum_{\gamma \varepsilon \Gamma} \gamma p_\gamma}$$

$$= \frac{\theta E[(H_X(t))^2 | H_X(t) \neq 0]}{E[H_X(t) | H_X(t) \neq 0]}$$

as $T \to \infty$, completing the proof.

In certain situations, such as birth-death processes in steady state, it may be true that $E[H_X(t) | H_X(t) \neq 0] = 0$, and hence, Theorem 6.1 may be of little value. However, in this case it may be possible to apply the following theorem.

<u>THEOREM 6.2</u>. Let the hypotheses of Lemma 6.1 be

satisfied, and let Γ be a finite set. If $N(T)/T \underset{p}{\to} \phi$ as $T \to \infty$, and $H_X(t)$, given $H_X(t) \neq 0$, has a finite second moment, then

$$\hat{\sigma}_T^2 \underset{p}{\to} \phi\, E[(H_X(t))^2 | H_X(t) \neq 0] \tag{6.9}$$

as $T \to \infty$.

Proof: Since $N(T) \underset{p}{\to} \infty$ as $T \to \infty$, given $\varepsilon > 0$, there exists $T_\varepsilon > 0$ such that

$$P[N(T) > 0] > 1 - \varepsilon$$

when $T > T_\varepsilon$. Then, if $T > T_\varepsilon$,

$$\hat{\sigma}_T^2 = \frac{N(T)}{T} \sum_{\gamma \varepsilon \Gamma} \gamma^2 \frac{N_\gamma(T)}{N(T)}$$

with probability greater than $1 - \varepsilon$. Since Γ is finite, by applying Lemma 6.1, we see that

$$\sum_{\gamma \varepsilon \Gamma} \gamma^2 \frac{N_\gamma(T)}{N(T)} \underset{p}{\to} E[(H_X(t))^2 | H_X(t) \neq 0]$$

as $T \to \infty$. Thus, since $N(T)/T \underset{p}{\to} \phi$ as $T \to \infty$, we have

$$\hat{\sigma}_T^2 \underset{p}{\to} \phi\, E[(H_X(t))^2 | H_X(t) \neq 0]$$

as $T \to \infty$.

If the hypotheses of Theorem 6.1 or Theorem 6.2 are satisfied we are now able to determine whether or not $\hat{\sigma}_T^2$ is a consistent estimator of σ^2 if we can compute the variance of $[\hat{\theta}(a,a+1)]$ and the right side of (6.8). Thus, the results of Theorem 6.1 and 6.2 can be summarized as follows. If the hypotheses of Theorem 6.1 are satisfied and

$$\text{var }[\hat{\theta}(a,a+1)] = \frac{\theta\ E[(H_X(t))^2 \mid H_X(t) \neq 0]}{E[H_X(t) \mid H_X(t) \neq 0]} ,$$

or if the hypotheses of Theorem 6.2 are satisfied and

$$\text{var }[\hat{\theta}(a,a+1)] = \phi\ E[(H_X(t))^2 \mid H_X(t) \neq 0] ,$$

then σ_T^2 is a consistent estimator of σ^2. These results will be illustrated in the following example.

Example 6.1. In order to obtain approximate confidence intervals for the reliability function

$$r(\lambda,x) = e^{-\lambda x}$$

associated with the Poisson process $\{N(t) \mid t \in [0,\infty)\}$, where $N(t)$ has parameter λt, we first note that in the setting of Example 2.1 each random variable of the process $\{H_N(t) | t \in [0,\infty)\}$ has the conditional distribution given by

$$P[H_N(t) = 1 \mid H_N(t) \neq 0] = 1 \quad .$$

Thus

$$E[H_N(t) \mid H_N(t) \neq 0] = 1$$

and

$$E[(H_N(t))^2 | H_N(t) \neq 0] = 1 \quad .$$

Also observe that since

$$\hat{\lambda} = \frac{N(T)}{T} \quad ,$$

we have

$$\hat{\lambda} \underset{p}{\rightarrow} \lambda$$

as $T \rightarrow \infty$. Therefore, since the hypotheses of Theorem 6.1 are satisfied, we obtain

$$\hat{\sigma}_T^2 \underset{p}{\rightarrow} \lambda \frac{E[(H_N(t))^2 | H_N(t) \neq 0]}{E[H_N(t) | H_N(t) \neq 0]} = \lambda$$

as $T \to \infty$. Also, $\hat{\lambda}(0,1)$ has the Poisson distribution with parameter λ, and hence,

$$S_T^2 \xrightarrow[p]{} \text{var}\,[\hat{\lambda}(0,1)]$$

as $T \to \infty$, so that Theorem 4.2 applies. Thus if we let

$$S_T^2 = \sum_{\gamma\varepsilon\Gamma} N_\gamma [r(\hat{\lambda} - \frac{\gamma}{T}, x) - r(\hat{\lambda}, x)]^2$$

$$= N(T)[e^{-(\hat{\lambda}-(1/T))x} - e^{-\hat{\lambda}x}]^2$$

$$= N(T)\, e^{-2\hat{\lambda}x} (e^{x/T} - 1)^2 ,$$

we have by Theorem 4.2

$$\frac{J_\infty(r(\hat{\lambda},x)) - r(\lambda,x)}{S_T} \xrightarrow[L]{} N(0,1)$$

as $T \to \infty$, and hence, the interval

$$[J_\infty(r(\hat{\lambda},x)) - S_T\, t_{\alpha/2}\, , J_\infty(r(\hat{\lambda},x)) + S_T\, t_{\alpha/2}] , \qquad (6.10)$$

where $t_{\alpha/2}$ is the $100(1 - \frac{\alpha}{2})$ percentile point of the $N(0,1)$ distribution, is an approximate

$100(1 - \alpha)\%$ confidence interval for $r(\lambda,x)$.

In order to exemplify this result the following Monte Carlo studies were made. Random Poisson numbers with known parameters were generated and substituted into (6.10). In Table 6.1, the columns labeled P_1, P_2, and P_3 give the percent of 1000 samples generated for which $r(\lambda,x)$ was contained in the confidence interval obtained from (6.10).

TABLE 6.1

λ	x	T	P_1	P_2	P_3
			$1 - \alpha = 0.50$	$1 - \alpha = 0.70$	$1 - \alpha = 0.95$
2	0.05	0.5	35.9	54.5	61.0
2	0.05	2.5	44.6	59.1	86.3
2	0.05	5.0	49.0	65.0	92.0
5	0.02	0.2	36.1	55.6	63.4
5	0.02	1.0	44.2	63.8	86.8
5	0.02	2.0	47.8	63.8	91.4
10	0.01	0.1	37.9	54.4	62.1
10	0.01	0.5	50.8	62.7	86.5
10	0.01	1.0	46.3	64.7	91.5

It is easily seen from Table 6.1 that when T is sufficiently large for the expected number of failures to be near 10, the values of P_1, P_2, and P_3

are quite close to $100(1 - \alpha)$. Thus, it appears that in this example one can apply Theorem 4.2 with confidence, even though T may be relatively small, provided that a sufficiently large number of failures can be expected.

CHAPTER V
THE $J_{\infty}^{(2)}$ ESTIMATOR

1. Introduction

In the previous chapter we introduced the notion of jackknifing stochastic processes. However, our study was limited to the counterparts of $J^{(1)}(\hat{\theta})$ and $G^{(1)}(\hat{\theta})$. In this chapter we will consider similar extensions for $J^{(2)}(\hat{\theta})$. Our approach will be essentially the same as in Chapter IV. However, more proofs of results will be included since, although the major results of this chapter were stated in [21], their proofs were not given. A more complete study of the $J^{(2)}$ estimator which includes the results of this chapter can be found in [1].

2. The $J_{\infty}^{(2)}$ Estimator

Let us assume the structure that was given in Chapter IV for the estimator $\hat{\theta}$ and let us also maintain the notation of that chapter. Then we make the following definition.

DEFINITION 2.1. For the regular partition $a = t_0 < t_1 < \ldots < t_n = b$ of the interval $[a,b]$ we define the $J^{(2)}$ estimator on a stochastic process by

$$J^{(2)}[f(\hat{\theta});n] = \frac{1}{2}\,[n^2 f(\hat{\theta}) - 2(n-1)^2\,\overline{f(\hat{\theta}_n^i)} + (n-2)^2\,\overline{f(\hat{\theta}_n^{ij})}\,] \quad , \qquad (2.1)$$

where

$$\hat{\theta}_n^{ij} = \frac{n\,\hat{\theta}}{n-2} - \frac{\hat{\theta}_i + \hat{\theta}_j}{n-2} \quad , \quad 1 \le i,j \le n, i \ne j \; , \qquad (2.2)$$

and

$$\overline{f(\hat{\theta}_n^{ij})} = \frac{1}{n(n-1)} \sum_{i \ne j}\sum f(\hat{\theta}_n^{ij}) \; . \qquad (2.3)$$

Although it is possible to establish theorems for $J^{(2)}[f(\hat{\theta});n]$ similar to those for $J[f(\hat{\theta});n]$ we will not pursue those for finite n but proceed directly to the limiting case.

DEFINITION 2.2. Let f be a function twice differentiable on the range of $\hat{\theta}$. Then we define the $J_\infty^{(2)}$ estimator by

$$J_\infty^{(2)}[f(\hat{\theta})] = f(\hat{\theta}) + \frac{1}{2} f''(\hat{\theta}) \left[\sum_{\gamma\varepsilon\Gamma} \frac{\gamma N_\gamma}{T} \right]^2$$

$$- \sum_{\gamma\varepsilon\Gamma} N_\gamma [(\tfrac{\gamma}{T})\;(N+1) f'(\hat{\theta}) - 2f(\hat{\theta})$$

$$+ 2f(\hat{\theta} - \tfrac{\gamma}{T})] + \frac{1}{2} \sum_{\gamma\varepsilon\Gamma} N_\gamma (N_\gamma - 1)[f(\hat{\theta})$$

$$- 2f(\hat{\theta} - \tfrac{\gamma}{T}) + f(\hat{\theta} - \tfrac{2\gamma}{T}) + (\tfrac{2\gamma}{T}) f'(\hat{\theta} - \tfrac{\gamma}{T})]$$

$$+ \frac{1}{2} \sum_{\gamma \neq \alpha}\sum N_\alpha N_\gamma [f(\hat{\theta}) - 2f(\hat{\theta} - \tfrac{\gamma}{T}) + f(\hat{\theta} - \frac{\gamma + \alpha}{T})$$

$$+ (\tfrac{2\gamma}{T}) f'(\hat{\theta} - \tfrac{\alpha}{T})] \quad , \qquad (2.4)$$

where

$$f''(\hat{\theta}) = \left.\frac{d^2 f(\theta)}{d\theta^2}\right|_{\theta = \hat{\theta}} \quad ,$$

$f'(\hat{\theta})$, Γ and N_γ are defined as in Definition IV.3.1, and

$$N = \sum_{\gamma \varepsilon \Gamma} N_\gamma \quad .$$

Ones first reaction to $J_\infty^{(2)}[f(\hat{\theta})]$ is possible regurgitation, or at least one of dismay. However, we hasten to point out that the $J_\infty^{(2)}$ estimator is not nearly as distasteful as it first appears. In fact in many cases it turns out to be every bit as manageable as the J_∞ estimator. Moreover in a sense (which we will discuss later) it is more robust than $J_\infty[f(\hat{\theta})]$ and possesses essentially the same asymptotic normality properties.

One should note that $J_\infty^{(2)}[f(\hat{\theta})]$ can be interpreted as $f(\hat{\theta})$ plus a correction term for bias and that $J_\infty[f(\hat{\theta})]$ has a similar interpretation. However, it is also interesting to note that $J_\infty^{(2)}[f(\hat{\theta})]$ can also be interpreted as $J_\infty[f(\hat{\theta})]$ plus a correction term for the bias in $J_\infty[f(\hat{\theta})]$. That is, by simple manipulation we obtain

$$
\begin{aligned}
J_\infty^{(2)}[f(\hat{\theta})] = J_\infty[f(\hat{\theta})] &+ \frac{1}{2} f''(\hat{\theta})\left[\sum_{\gamma\varepsilon\Gamma} \frac{\gamma N_\gamma}{T}\right]^2 \\
&- \sum_{\gamma\varepsilon\Gamma} N_\gamma \left[(\frac{\gamma}{T}) N f'(\hat{\theta}) - f(\hat{\theta}) \right. \\
&\left. + f(\hat{\theta} - \frac{\gamma}{T})\right] \\
&+ \frac{1}{2} \sum_{\gamma\varepsilon\Gamma} N_\gamma (N_\gamma - 1)[f(\hat{\theta}) - 2f(\hat{\theta} - \frac{\gamma}{T}) \\
&+ f(\hat{\theta} - \frac{2\gamma}{T}) + (\frac{2\gamma}{T}) f'(\hat{\theta} - \frac{\gamma}{T})] \\
&+ \frac{1}{2} \sum_{\gamma \neq \alpha}\sum N_\gamma N_\alpha [f(\hat{\theta}) - 2f(\hat{\theta} - \frac{\gamma}{T}) \\
&+ f(\hat{\theta} - \frac{\gamma + \alpha}{T}) + (\frac{2\gamma}{T}) f'(\hat{\theta} - \frac{\alpha}{T})] \quad .
\end{aligned}
\tag{2.5}
$$

Just as $J_\infty[f(\hat{\theta})]$ simplified, $J_\infty^{(2)}[f(\hat{\theta})]$ admits of several simplifications which will now be listed as examples paralleling Examples 3.1 through 3.4 of Chapter IV.

Example 2.1. Suppose that with probability one, every realization of $\{I_X(t) | t \varepsilon [a,b]\}$ is a step function, then

$$\hat{\theta} = \sum_{\gamma \varepsilon \Gamma} \frac{\gamma N_\gamma}{T} ,$$

and therefore

$$J_\infty^{(2)}[f(\hat{\theta})] = f(\hat{\theta}) + \frac{\hat{\theta}^2 f''(\hat{\theta})}{2} - (N+1)\,\hat{\theta} f'(\hat{\theta})$$

$$+ 2 \sum_{\gamma \varepsilon \Gamma} N_\gamma \left[f(\hat{\theta}) - f\left(\hat{\theta} - \frac{\gamma}{T}\right)\right]$$

$$+ \frac{1}{2} \sum_{\gamma \varepsilon \Gamma} N_\gamma (N_\gamma - 1)\left[f(\hat{\theta}) - 2f\left(\hat{\theta} - \frac{\gamma}{T}\right)\right.$$

$$\left. + f\left(\hat{\theta} - \frac{2\gamma}{T}\right) + \left(\frac{2\gamma}{T}\right) f'\left(\hat{\theta} - \frac{\gamma}{T}\right)\right]$$

$$+ \frac{1}{2} \sum_{\gamma \neq \alpha}\sum N_\gamma N_\alpha \left[f(\hat{\theta}) - 2f\left(\hat{\theta} - \frac{\gamma}{T}\right)\right.$$

$$\left. + f\left(\hat{\theta} - \frac{\alpha + \gamma}{T}\right) + \left(\frac{2\gamma}{T}\right) f'\left(\hat{\theta} - \frac{\alpha}{T}\right)\right] . \tag{2.6}$$

Example 2.2. If $\Gamma = \{\gamma_0\}$, i.e., all discontinuities are of the same size, then the double summation in (2.5) is eliminated and we have

$$J_\infty^{(2)}[f(\hat{\theta})] = f(\hat{\theta}) + [N^2 f''(\hat{\theta})\gamma_0^2]/[2T^2]$$

$$- N[(\frac{\gamma_0}{T})(N+1)f'(\hat{\theta}) - 2f(\hat{\theta})$$

$$+ 2f(\hat{\theta} - \frac{\gamma_0}{T})] + \frac{N(N-1)}{2}[f(\hat{\theta}) - 2f(\hat{\theta} - \frac{\gamma_0}{T})$$

$$+ f(\hat{\theta} - \frac{2\gamma_0}{T}) + (\frac{2\gamma_0}{T})f'(\hat{\theta} - \frac{\gamma_0}{T})]. \qquad (2.7)$$

Example 2.3. If the assumptions of both Examples 2.1 and 2.2 hold, then $J_{\infty}^{(2)}[f(\hat{\theta})]$ simplifies further and we have

$$J_{\infty}^{(2)}[f(\hat{\theta})] = f(\hat{\theta}) + \frac{\hat{\theta}^2 f''(\hat{\theta})}{2} - (N+1)\ \hat{\theta}f'(\hat{\theta})$$

$$+ 2N[f(\hat{\theta}) - f(\hat{\theta} - \frac{\gamma_0}{T})]$$

$$+ \frac{N(N-1)}{2}[f(\hat{\theta}) - 2f(\hat{\theta} - \frac{\gamma_0}{T})$$

$$+ f(\hat{\theta} - \frac{2\gamma_0}{T}) + (\frac{2\gamma_0}{T})\ f'(\hat{\theta} - \frac{\gamma_0}{T})]\ .$$

$$(2.8)$$

Example 2.4. Finally if $\Gamma = \emptyset$, the empty set, i.e., if the process $\{I_X(t) | t \in [a,b]\}$ is continuous with probability one, we have

$$J_{\infty}^{(2)}[f(\hat{\theta})] = J_{\infty}[f(\hat{\theta})] = f(\hat{\theta}) \quad . \qquad (2.9)$$

We shall now consider a simple example from [21] to provide a specific illustration of the estimator of Definition 2.2.

Example 2.5. Let $\{N(t) | t \in [a,b]\}$ be the Poisson process once more. Let $\hat{\lambda} = N/T$ and let $f(\hat{\lambda}) = (N/T)^3$ be an estimator for λ^3. Then

$$J_\infty[f(\hat{\theta})] = (\frac{N}{T})^3 - \frac{3N^2}{T^3} + \frac{N}{T^3}$$

and

$$J_\infty^{(2)}[f(\hat{\theta})] = (\frac{N}{T})^3 - \frac{3N^2}{T^3} + \frac{2N}{T^3} \quad .$$

Moreover it is easy to show that

$$E[J_\infty(f(\hat{\theta}))] = \lambda^3 - \frac{\lambda}{T^2}$$

and

$$E[J_\infty^{(2)}(f(\hat{\theta}))] = \lambda^3 \quad .$$

Consequently, since $\hat{\lambda}$ is a complete sufficient statistic for λ, $J^{(2)}[f(\hat{\theta})]$ is the unique minimum variance unbiased estimator of λ^3. Moreover, one can also show in this case that when $\lambda, T \geq 1$, the MSE $J_\infty^{(2)}[f(\hat{\theta})] \leq$ MSE $J_\infty[f(\hat{\theta})] \leq$ MSE $f(\hat{\theta})$.

This rather trivial example suggests that in many cases $J_\infty^{(2)}[f(\hat{\theta})]$ may be no more difficult to manage than $J_\infty[f(\hat{\theta})]$ and in some of these cases it will be the more desirable of the two estimators. Thus the $J_\infty^{(2)}$ estimator does appear to be of sufficient interest to warrant our further consideration.

The next theorem establishes that $J_\infty^{(2)}[f(\hat{\theta})]$ is indeed the limiting estimator associated with $J^{(2)}[f(\hat{\theta});n]$.

THEOREM 2.1. Suppose the conditions of Theorem IV.3.1 are satisfied and that f has a continuous second derivative in a neighborhood of θ. Then

$$\lim_{n\to\infty} J^{(2)}[f(\hat{\theta});n] = J_\infty^{(2)}[f(\hat{\theta})] \tag{2.10}$$

with probability one.

The proof of this result is rather lengthy and was first given in [1]. Since that document is not generally available, for the sake of completeness, the proof from [1] is given in the Appendix with the author's permission.

Since $J_\infty^{(2)}[f(\hat{\theta})]$ was obtained in essentially the same manner as $J_\infty[f(\hat{\theta})]$, one would expect it to have similar properties. This is the case and we shall devote the next few sections to establishing these properties.

3. Bias Reduction Properties

Although it is not necessary to assume the process $\{I_X(t)|t \in [a,b]\}$ has stationary independent increments in order to show $J^{(2)}[f(\hat{\theta});n] \to J_\infty[f(\hat{\theta})]$ as $n \to \infty$, it is essentially necessary to make this assumption to obtain any useful results regarding bias reduction. That is, just as in the case of $J_\infty[f(\hat{\theta})]$, we will find it helpful in establishing the bias reduction properties of $J_\infty^{(2)}[f(\hat{\theta})]$ to assume that the bias in $f(\hat{\theta}_n^i)$ is independent of which subinterval is deleted (recall Corollary IV.3.1). The only practical condition which guarantees this assumption is, of course, the assumption of stationary independent increments on the $\{I_X(t)\}$ process.

We now state and sketch a proof of the primary result regarding the bias reduction properties of $J_\infty^{(2)}[f(\hat{\theta})]$.

<u>THEOREM 3.1</u>. For $T > T_0$ let

$$E[f(\hat{\theta})] = f(\theta) + B(T,\theta)$$

$$E[f(\hat{\theta}^i)] = f(\theta) + B_{i,n}(T,\theta)$$

$$E[f(\hat{\theta}^{ij})] = f(\theta) + B_{ij,n}(T,\theta) \quad .$$

If

(i) $\lim\limits_{n\to\infty} E[J^{(2)}(f(\hat{\theta});n)] = E[J_\infty^{(2)}(f(\hat{\theta}))]$,

(ii) $B_{i,n}(T,\theta) = B_{j,n}(T,\theta)$ and $B_{ij,n}(T,\theta)$

$= B_{k\ell,n}(T,\theta)$ for all i,j,k, and ℓ ,

(iii) $\dfrac{\partial^2 B(T,\theta)}{\partial T^2}$ exists for $T > T_0$,

then

$$E[J_\infty^{(2)}(f(\hat{\theta}))] = f(\theta) + B(T,\theta) + \frac{2T\partial B(T,\theta)}{\partial T} + \frac{T^2}{2}\,\frac{\partial^2 B(\theta,T)}{\partial T^2} \quad , \qquad (3.1)$$

for all $T > T_0$.

Proof: By simple algebra one can verify that

$$J^{(2)}[f(\hat{\theta});n] = \frac{n}{2(n-1)} \sum_{i \neq j}\sum [f(\hat{\theta}) - 2f(\hat{\theta}^i) + f(\hat{\theta}^{ij})] + \frac{2}{n} \sum_{i \neq j}\sum [f(\hat{\theta}^i) - f(\hat{\theta}^{ij})] + \frac{1}{n}\sum_{i=1}^{n} f(\hat{\theta}^i) \quad .$$

Then

$$E[J^{(2)}(f(\hat{\theta};n)] = f(\theta) + \frac{n}{2(n-1)} \sum_{i \neq j}\sum [B(T,\theta) - 2B_{i,n}(T,\theta) + B_{ij,n}(T,\theta)]$$

$$+ \frac{2}{n} \sum_{i \neq j}\sum [B_{i,n}(T,\theta) - B_{ij,n}(T,\theta)]$$

$$+ \frac{1}{n} \sum_{i=1}^{n} B_{i,n}(T,\theta) \quad . \qquad (3.2)$$

But

$$B_{n,n}(T,\theta) = B(\frac{n-1}{n} T,\theta)$$

and

$$B_{(n-1)n,n}(T,\theta) = B(\frac{n-2}{n} T,\theta) \quad ,$$

and using (ii), we can therefore write (3.2) as follows:

$$E[J^{(2)}(f(\hat{\theta});n)] = f(\theta) + \frac{n^2}{2} [B(T,\theta) - 2B(\frac{n-1}{n} T,\theta)$$

$$+ B(\frac{n-2}{n} T,\theta)]$$

$$+ 2(n-1)[B(\frac{n-1}{n} T,\theta)$$

$$- B(\frac{n-2}{n} T,\theta)] + B(\frac{n-1}{n} T,\theta) \quad .$$

Then by (i) and the continuity of B we have

$$E[J_\infty^{(2)}(f(\hat{\theta}))] = f(\theta) + B(T,\theta)$$

$$+ 2 \lim_{n\to\infty} \{(n - 1)[B(\frac{n - 1}{n} T,\theta)$$

$$- B(\frac{n - 2}{n} T,\theta)]\}$$

$$+ \frac{1}{2} \lim_{n\to\infty} \{n^2[B(T,\theta) - 2B(\frac{n - 1}{n} T,\theta)$$

$$+ B(\frac{n - 2}{n} T,\theta)]\} \quad . \qquad (3.3)$$

Now letting $\Delta T = T/n$ we have

$$(n - 1)[B(\frac{n - 1}{n} T,\theta) - B(\frac{n - 2}{n} T,\theta)]$$

$$= - (n - 1)[B(T,\theta) - B(T - \Delta T,\theta) - (B(T,\theta)$$

$$+ B(T - 2 \Delta T,\theta))]$$

$$= - \frac{(n - 1)T}{n} \left[\frac{B(T,\theta) - B(T - \Delta T,\theta)}{\Delta T}\right]$$

$$+ \frac{2(n-1)T}{n}\left[\frac{B(T,\theta) - B(T - 2\,\Delta T,\theta)}{2\Delta T}\right] .$$

Thus

$$\lim_{n\to\infty} \{(n-1)[B(\frac{n-1}{n}T,\theta) - B(\frac{n-2}{n}T,\theta)]\}$$

$$= T\,\frac{\partial B(T,\theta)}{\partial T} . \qquad (3.4)$$

Moreover by writing

$$n^2[B(T,\theta) - 2B(\frac{n-1}{n}T,\theta) + B(\frac{n-2}{n}T,\theta)]$$

$$= T^2\left[\frac{B(T,\theta) - 2B(T - \Delta T,\theta) + B(T - 2\,\Delta T,\theta)}{(\Delta T)^2}\right] ,$$

it is clear that

$$\lim_{n\to\infty} \{n^2[B(T,\theta) - 2B(\frac{n-1}{n}T,\theta) + B(\frac{n-2}{n}T,\theta)]\}$$

$$= T^2\,\frac{\partial^2 B(T,\theta)}{\partial T^2} . \qquad (3.5)$$

Combining (3.3), (3.4), and (3.5) yields the desired result.

We should note, before proceeding to our next theorem, that condition (ii) of Theorem 3.1 could be replaced by the condition that $\{I_X(t) \mid t \in [a,b]\}$ be a stationary independent increment process and the theorem would remain valid.

Theorem 3.1 essentially characterized the bias in $J_\infty^{(2)}[f(\hat{\theta})]$ for stationary independent increment processes and yields the following important result which demonstrates the sense in which one can consider $J_\infty^{(2)}[f(\hat{\theta})]$ an extension of $J_\infty[f(\hat{\theta})]$.

<u>THEOREM 3.2.</u> Under the conditions of Theorem 3.1, $J_\infty^{(2)}[f(\hat{\theta})]$ is unbiased if and only if for $T > T_0$

$$B(T,\theta) = \frac{C_1(\theta)}{T} + \frac{C_2(\theta)}{T^2}, \tag{3.6}$$

where C_1 and C_2 are arbitrary functions of θ. Moreover, if

$$B(T,\theta) = \sum_{k=1}^{\infty} \frac{C_k(\theta)}{T^k}, \quad T > T_0 \geq 0,$$

then

$$E[J_\infty^{(2)}(f(\hat{\theta})) - f(\theta)] = \sum_{k=3}^{\infty} \frac{a_k(\theta)}{T^k}, \quad T > T_0, \tag{3.7}$$

where the $a_k(\theta)$ are determined by the $C_k(\theta)$.

Proof: The theorem is an immediate consequence of equation (3.1). That is,

$$B(T,\theta) + 2T \frac{\partial B(T,\theta)}{\partial T} + \frac{T^2}{2} \frac{\partial^2 B(\theta,T)}{\partial T^2} = 0, \quad T > T_0 \quad ,$$

if and only if (3.6) holds; and since (3.6) implies the validity of (3.7), the theorem follows.

As one might expect, when Theorem 3.1 obtains, it is possible to parallel the development of the bias reduction properties of $J(\hat{\theta})$ and $J_\infty[f(\hat{\theta})]$ studied in previous chapters. Since these theorems are useful in understanding the nature of $J_\infty^{(2)}[f(\hat{\theta})]$ we will now proceed with that development. However, since the interpretation of the results we are about to give follows in essentially the same way as our previous discussion, we will be brief.

THEOREM 3.3. Let the conditions of Theorem 3.1 be satisfied and suppose there exists a $p > 0$ such that

$$\lim_{T\to\infty} T^p B(T,\theta) = C(\theta) \neq 0, \pm \infty,$$

and

$$\lim_{T\to\infty} \left[T^{p+2} \frac{\partial^2 B(T,\theta)}{\partial T^2} \right]$$

exists.

Then

(i) if $p = 2$ or $p = 1$, then $J_\infty^{(2)}[f(\hat{\theta})]$ L.O.B.E. $f(\hat{\theta})$,

(ii) if $p < 3$, $p \neq 2$, $p \neq 1$, then

$J_\infty^{(2)}[f(\hat{\theta})]$ B.S.O.B.E. $f(\hat{\theta})$,

(iii) if $p = 3$, then $J_\infty^{(2)}[f(\hat{\theta})]$ S.O.B.E. $f(\hat{\theta})$,

(iv) if $p > 3$, then $f(\hat{\theta})$ B.S.O.B.E. $J_\infty^{(2)}[f(\hat{\theta})]$.

Proof:

$$\left| \lim_{T\to\infty} \frac{B(T,\theta) + 2T[\partial B(T,\theta)/\partial T] + (T^2/2)[\partial^2 B(T,\theta)/\partial T^2]}{B(T,\theta)} \right|$$

$$= \left| \lim_{T\to\infty} \frac{T^p B(T,\theta) + 2T^{p+1}[\partial B(T,\theta)/\partial T]}{T^p B(T,\theta)} + \frac{(T^{p+2}/2)[\partial^2 B(T,\theta)/\partial T^2]}{T^p B(T,\theta)} \right|$$

$$= \left| 1 + \frac{\lim_{T\to\infty} 2T^{p+1}[\partial B(T,\theta)/\partial T]}{C(\theta)} \right.$$

$$+ \frac{\lim_{T\to\infty} (T^{p+2}/2)[\partial^2 B(T,\theta)/\partial T^2]}{C(\theta)} \Bigg| \quad . \tag{3.8}$$

Now since $\lim_{T\to\infty} T^p B(T,\theta) = C(\theta) \neq 0$, then $\lim_{T\to\infty} B(T,\theta) = 0$. Thus using L'Hospital's rule

$$C(\theta) = \lim_{T\to\infty} T^p B(T,\theta) = \lim_{T\to\infty} \frac{B(T,\theta)}{T^{-p}}$$

$$\lim_{T\to\infty} \frac{[\partial B(T,\theta)/\partial T]}{-pt^{-p-1}} = \lim_{T\to\infty} \frac{[\partial^2 B(T,\theta)/\partial T]}{p(p+1)T^{-p-2}} \quad .$$

It now follows that $\lim_{T\to\infty} \frac{\partial B(T,\theta)}{\partial T} \cdot (T^{p+1}) = pC(\theta)$

and

$$\lim_{T\to\infty} \frac{T^{p+2}}{2} \frac{\partial^2 B(T,\theta)}{\partial T^2} = \frac{p(p+1)}{2} C(\theta) \quad .$$

Hence (3.8) becomes

$$\left| 1 + \frac{2(-pC(\theta))}{C(\theta)} + \frac{p(p+1)C(\theta)}{2C(\theta)} \right| = \frac{1}{2} \left| (p-2)(p-1) \right| \quad .$$

It is now easy to deduce the results (i), (ii), (iii), (iv) from the above expression.

A result similar to Theorem 3.3 can be obtained for comparing $J_\infty[f(\hat{\theta})]$ with $J_\infty^{(2)}[f(\hat{\theta})]$. In order to accomplish this it is, however, necessary to consider two separate cases, namely $p \neq 1$ and $p = 1$. This is the subject of the next two theorems.

<u>THEOREM 3.4</u>. Suppose

$$E[J_\infty[f(\hat{\theta})]] = f(\theta) + B(T,\theta) + T\,\frac{\partial B(T,\theta)}{\partial T}$$

and

$$E[J_\infty^{(2)}[f(\hat{\theta})]] = f(\theta) + B(T,\theta) + 2T\,\frac{\partial B(T,\theta)}{\partial T} + \frac{T^2}{2}\,\frac{\partial^2 B(T,\theta)}{\partial T}$$

and further that there exists $p > 0$, $p \neq 1$ such that

$$\lim_{T\to\infty} T^p B(T,\theta) = C(\theta) \neq 0, \pm\infty$$

and $\lim_{T\to\infty} T^{p+2}[\partial^2 B(T,\theta)/\partial T^2]$ exists. Then

(i) if $p = 2$, $J_\infty^{(2)}[f(\hat{\theta})]$ L.O.B.E. $J_\infty[f(\hat{\theta})]$,

(ii) if $p < 4$, $p \neq 2$, $J_{\infty}^{(2)}[f(\hat{\theta})]$ B.S.O.B.E. $J_{\infty}[f(\hat{\theta})]$,

(iii) if $p = 4$, $J_{\infty}^{(2)}[f(\hat{\theta})]$ S.O.B.E. $J_{\infty}[f(\hat{\theta})]$,

(iv) if $p > 4$, $J_{\infty}[f(\hat{\theta})]$ B.S.O.B.E. $J_{\infty}^{(2)}[f(\hat{\theta})]$.

Proof: Since $p > 0$ and $p \neq 1$, $J_{\infty}[f(\hat{\theta})]$ is not unbiased over any interval and we can write

$$\left| \lim_{T\to\infty} \frac{B(T,\theta) + 2T[\partial B(T,\theta)/\partial T] + (T^2/2)[\partial^2 B(T,\theta)/\partial T^2]}{B(T,\theta) + T[\partial B(T,\theta)/\partial T]} \right|$$

$$= \left| 1 + \lim_{T\to\infty} \frac{T^{p+1}[\partial B(T,\theta)/\partial T] + (T^{p+2}/2)[\partial^2 B(T,\theta)/\partial T^2]}{T^p B(T,\theta) + T^{p+1}[\partial B(T,\theta)/\partial T]} \right|$$

$$= \left| 1 + \frac{-pC(\theta) + \frac{1}{2}p(p+1)C(\theta)}{C(\theta) - pC(\theta)} \right|$$

$$= \left| 1 - \frac{1}{2}p \right| \quad , \tag{3.9}$$

The theorem then easily follows by inspection of (3.9).

<u>THEOREM 3.5</u>. Suppose that conditions of Theorem 3.4

are satisfied except that p = 1. If there exists a $p_1 > 0$ such that

$$\lim_{T\to\infty} T^{p_1}[TB(T,\theta) - C(\theta)] = C_1(\theta) \neq 0 \ , \quad \pm\infty,$$

and lim as $t \to \infty$ of $T^{p_1+2}[\partial^2[TB(T,\theta)]/\partial T^2]$ exists, then

(i) if $p_1 = 1$, $J_\infty^{(2)}[f(\hat{\theta})]$ L.O.B.E. $J_\infty[f(\hat{\theta})]$,

(ii) if $0 < p_1 < 3$, $p_1 \neq 1$, $J_\infty^{(2)}[f(\hat{\theta})]$ B.S.O.B.E. $J_\infty[f(\hat{\theta})]$,

(iii) if $p_1 = 3$, $J_\infty^{(2)}[f(\hat{\theta})]$ S.O.B.E. $J_\infty[f(\hat{\theta})]$,

(iv) if $p_1 > 3$, $J_\infty[f(\hat{\theta})]$ B.S.O.B.E. $J_\infty^{(2)}[f(\hat{\theta})]$.

Proof: By L'Hospital's rule we have

$$C_1(\theta) = \lim_{T\to\infty} \frac{TB(T,\theta) - C(\theta)}{T^{-p_1}}$$

$$= \lim_{T\to\infty} \frac{B(T,\theta) + T[\partial B(T,\theta)/\partial T]}{-p_1 T^{-p_1-1}}$$

$$= \lim_{T\to\infty} \frac{2[\partial B(T,\theta)/\partial T] + T[\partial^2 B(T,\theta)/\partial T^2]}{p_1(p_1 + 1)T^{-p_1-2}}$$

$$= \lim_{T\to\infty} T^{p_1+2} \frac{\partial^2[TB(T,\theta)]}{\partial T^2} \cdot \frac{1}{p_1(p_1 + 1)} . \quad (3.10)$$

Therefore

$$\lim_{T\to\infty} T^{p_1+1} [B(T,\theta) + T \frac{\partial B(T,\theta)}{\partial T}] = - p_1 C_1(\theta)$$

and

$$\lim_{T\to\infty} T^{p_1+2} \left[2 \frac{\partial B(T,\theta)}{\partial T} + T \frac{\partial^2 B(T,\theta)}{\partial T^2} \right]$$

$$= p_1(p_1 + 1) C_1(\theta) .$$

Now consider

$$\left|\lim_{T\to\infty} \frac{B(T,\theta) + 2T[\partial B(T,\theta)/\partial T] + (T^2/2)[\partial^2 B(T,\theta)/\partial T^2]}{B(T,\theta) + T[\partial B(T,\theta)/\partial T]}\right|$$

$$= \frac{1}{2}\left|2 + \lim_{T\to\infty} \frac{T^{p_1+1}[2T[\partial B(T,\theta)/\partial T] + T^2\ \partial^2 B(T,\theta)/\partial T^2]}{T^{p_1+1}[B(T,\theta) + T\ \partial B(T,\theta)/\partial T]}\right|$$

$$= \frac{1}{2}\left|2 + \frac{p_1(p_1 + 1)C_1(\theta)}{-p_1\ C_1(\theta)}\right|$$

$$= \frac{1}{2}\left|p_1 - 1\right| \quad .$$

The results (i), (ii), (iii), and (iv) now easily follow.

<u>Example 3.1</u>. Consider once more our previous example where $\{N(t)|t \in [0,\infty)\}$ is a Poisson process with failure rate λ. Then, again, let $r(\hat{\lambda};x) = \exp(-\hat{\lambda} x)$, where $\hat{\lambda} = N(T)/T$. We shall find the bias in $J_\infty[r(\hat{\lambda};x)]$ and $J_\infty^{(2)}[r(\hat{\lambda};x)]$ and show that $J_\infty^{(2)}[r(\hat{\lambda};x)]$ L.O.B.E. $J_\infty[r(\hat{\lambda};x)]$ L.O.B.E. $r(\hat{\lambda};x)$.

In Example IV.3.7. we showed that

$$B(T,\lambda) = \exp[-\lambda T(1 - \exp(-\tfrac{x}{T}))] - \exp(-\lambda x)$$

and $J_\infty[r(\hat{\lambda};x)]$ L.O.B.E. $r(\hat{\lambda};x)$. Moreover in that example we used the result for the J_∞ estimator which corresponds to Theorem 3.1 to show that

$$E[J_\infty(r(\hat{\lambda};x))] - \exp(-\lambda x)$$

$$= \{\exp[-\lambda T(1 - \exp(-\tfrac{x}{T}))]\}\{[1 - \lambda T$$

$$+ \lambda T(1 + \tfrac{x}{T}) \exp(-\tfrac{x}{T})]\} - \exp(-\lambda x) \quad .$$

Further, by showing

$$\lim_{T\to\infty} TB(T,\lambda) = \frac{1}{2}\lambda x^2 e^{-\lambda x} \neq 0 \quad ,$$

we established that $J_\infty[r(\hat{\lambda};x)]$ L.O.B.E. $r(\hat{\lambda};x)$. At this point, with the aid of Theorems 3.1 and 3.5, we can now find an expression for the bias in $J_\infty^{(2)}[r(\hat{\lambda};x)]$ and show that $J_\infty^{(2)}[r(\hat{\lambda};x)]$ L.O.B.E. $J_\infty[r(\hat{\lambda};x)]$. That is, by Theorem 3.1

$$E[J_\infty^{(2)}(r(\hat{\lambda};x))] - r(\lambda;x)$$

$$= B(T,\lambda) + 2T \frac{\partial B(T,\lambda)}{\partial T} + \frac{T^2}{2} \frac{\partial^2 B(T,\lambda)}{\partial T^2}$$

$$= \{\exp[-\lambda T(1 - \exp(-\tfrac{x}{T}))]\}\{1$$

$$- 2\lambda T(1 - \tfrac{x}{T}\exp(-\tfrac{x}{T}) - \exp(-\tfrac{x}{T})]$$

$$+ \frac{(\lambda T)^2}{2}[[1 - (\tfrac{x}{T} + 1)\exp(-\tfrac{x}{T})]^2$$

$$+ \left[\frac{x^2}{T^3} + \frac{2x}{T^2}\right]\exp(-\tfrac{x}{T})]\} - \exp(-\lambda x) .$$

To show that $J_\infty^{(2)}[r(\hat{\lambda};x)]$ L.O.B.E. $J_\infty[r(\hat{\lambda};x)]$ we note that

$$T[T(B(T,\lambda) - C(\theta)] = T\left[T(B(T,\lambda)) - \frac{\lambda x^2 \exp(-\lambda x)}{2}\right]$$

and hence (with a little effort)

$$\lim_{T\to\infty} T[T(B(T,\lambda)) - C(\theta)] = \frac{-\lambda x^3}{6} \neq 0 .$$

From Theorem 3.5 it therefore follows that $J_\infty^{(2)}[f(\hat{\theta})]$ L.O.B.E. $J_\infty[f(\hat{\theta})]$. Before leaving this section on bias reduction we should point out that, as in the case of $J^{(k)}[\hat{\theta}]$, there is a class of distributions for which unbiased estimators can be completely characterized through $J_\infty^{(k)}[f(\hat{\theta})]$. The following theorem, which is analogous to Theorem I.5.6, establishes this result.

<u>THEOREM 3.6</u>. Let $\{I_X(t)|t \varepsilon [a,\infty)\}$ be a stationary independent increment process and suppose that $[\partial^2 B(T,\theta)/\partial T^2]$ exists and $\lim_{n\to\infty} E[J^{(k)}(f(\hat{\theta});n)] = E[J_\infty^{(k)}[f(\hat{\theta})]]$, $k = 0,1,2$, where $J_\infty^{(0)}[f(\hat{\theta})]$ and $J_\infty^{(1)}[f(\hat{\theta})]$ denote $f(\hat{\theta})$ and $J_\infty[f(\hat{\theta})]$, respectively. Moreover, suppose that $f(\hat{\theta})$ is asymptotically unbiased, i.e.,

$$\lim_{T\to\infty} E[f[\hat{\theta}(a, a + T)]] = f(\theta) \quad .$$

Now, if $k = 0$ or 1 and

$$J_\infty^{(k+1)}[f(\hat{\theta})] = J_\infty^{(k)}[f(\hat{\theta})] \quad ,$$

for $T > T_0$, then $J_\infty^{(k)}[f(\hat{\theta})]$ is an unbiased estimator for $f(\theta)$, for all $T > T_0$. Further, if for $T > T_0$

$$J_\infty^{(k+1)}[f(\hat{\theta})] \not\equiv J_\infty^{(k)}[f(\hat{\theta})] \tag{3.12}$$

and $J_\infty^{(k+1)}[f(\hat{\theta})] - J_\infty^{(k)}[f(\hat{\theta})]$ is complete, then $J_\infty^{(k)}[f(\hat{\theta})]$ is a biased estimator for $f(\theta)$.

Proof:

Case 1: $k = 0$.

Suppose (3.11) holds, i.e., $J_\infty[f(\hat{\theta})] = f(\hat{\theta})$, when $T > T_0$. Then

$$B(T,\theta) + T \frac{\partial B(T,\theta)}{\partial T} = B(T,\theta)$$

for all $T > T_0$ and hence

$$T \frac{\partial B(T,\theta)}{\partial T} = 0 \qquad (3.13)$$

when $T > T_0$. But equation (3.13), subject to the boundary condition that $\lim_{T\to\infty} B(T,\theta) = 0$, has the solution $B(T,\theta) = 0$. Thus, $f(\hat{\theta})$ is unbiased. Now suppose (3.12) holds and $f(\hat{\theta})$ is unbiased. Clearly, if $f(\hat{\theta})$ is unbiased $J_\infty[f(\hat{\theta})]$ is unbiased by Corollary IV.3.1 and hence

$$E[J_\infty(f(\hat{\theta})) - f(\hat{\theta})] = 0 \qquad (3.14)$$

for all $T > T_0$ and all θ. However the completeness of the difference in (3.14) implies

$$J_\infty(f(\hat{\theta})) = f(\hat{\theta}) \quad ,$$

and the proof of Case 1 is established by contradiction.

Case 2: $k = 1$.

Again suppose equation (3.11) holds, that is, $J_\infty^{(2)}(f(\hat{\theta})) = J_\infty[f(\hat{\theta})]$ when $T > T_0$. Then

$$B(T,\theta) + 2T \frac{\partial B(T,\theta)}{\partial T} + \frac{T^2}{2} \frac{\partial^2 B(T,\theta)}{\partial T^2}$$

$$= B(T,\theta) + T \frac{\partial B(T,\theta)}{\partial T} , \qquad (3.15)$$

or

$$\frac{\partial B(T,\theta)}{\partial T} + \frac{T}{2} \frac{\partial^2 B(T,\theta)}{\partial T^2} = 0 \qquad (3.16)$$

when $T > T_0$. But equation (3.16) and the condition $B(T,\theta) \to 0$ as $t \to \infty$ is satisfied if and only if $B(T,\theta) = C(\theta)/T$, where $C(\theta)$ is an arbitrary function of θ. Therefore $J_\infty[f(\hat{\theta})]$ is unbiased. The remaining part of the proof follows by noting that $J_\infty[f(\hat{\theta})]$ unbiased implies that $J_\infty^{(2)}[f(\hat{\theta})]$ is also unbiased and then by proceeding exactly as in Case 1.

In concluding our discussion on the $J_\infty^{(2)}$ estimator we include one final important section in which it is shown that this estimator possesses essentially the same asymptotic distribution properties as $J_\infty[f(\hat{\theta})]$. Since it often has better bias reduction properties than $J_\infty[f(\hat{\theta})]$, these latter results suggest that some additional study of both the $J^{(2)}$ and $J_\infty^{(2)}$ estimators would be an interesting research project. The authors are indebted to Professors J. E. Adams and T. A. Watkins who have generously contributed this final section.

4. Asymptotic Distribution Properties of the $J_\infty^{(2)}$ Estimator (J. E. Adams and T. A. Watkins)

In this section we will show that as the length of the interval, $T = b - a$, increases without bound, the limiting distribution of $J_\infty^{(2)}[f(\hat{\theta})]$ is the same as that of $J_\infty[f(\hat{\theta})]$. This will enable one to test hypotheses and obtain approximate confidence intervals for $f(\theta)$ using $J_\infty^{(2)}[f(\hat{\theta})]$. If $J_\infty^{(2)}[f(\hat{\theta})]$ has smaller bias than $J_\infty[f(\hat{\theta})]$, these confidence intervals may be better centered about the parameter $f(\theta)$. Although it is not always evident in the notation, the dependence of $\hat{\theta}$, N, N_γ , etc. on T will be tacitly assumed as in Section 4 of Chapter IV.

<u>THEOREM 4.1</u>. Let $\{I_X(t) | t \in [a,\infty)\}$ be a stochastic process with stationary independent increments such that $\text{var}[\hat{\theta}(a, a+1)] = \sigma^2 < \infty$ and $E[\hat{\theta}(a,t)] = \theta$ for every $t > a$. Also let f be a function such that f''' exists and is bounded in a neighborhood of θ. If $T^{-5/4}N \underset{p}{\to} 0$ as $T \to \infty$, and if Γ is a bounded set, then

$$\sqrt{T}\,[J_\infty^{(2)}[f(\hat{\theta})] - f(\theta)] \underset{\mathcal{L}}{\to} N(0,\sigma^2[f'(\theta)]^2)$$

as $T \to \infty$.

Proof: Under the hypotheses of this theorem, it was shown in [48] that

$$\sqrt{T}\,[f(\hat{\theta}) - f(\theta)] \underset{\mathcal{L}}{\rightarrow} N(0,\sigma^2[f'(\theta)]^2)$$

as $T \rightarrow \infty$. In view of this result, it is sufficient to show that

$$\sqrt{T}\,[J_\infty^{(2)}[f(\hat{\theta})] - f(\hat{\theta})] \underset{p}{\rightarrow} 0 \tag{4.1}$$

as $T \rightarrow \infty$. By rearranging the terms of $J_\infty^{(2)}[f(\hat{\theta})]$ in Definition 1.2, we obtain

$$\begin{aligned} J_\infty^{(2)}[f(\hat{\theta})] = {} & f(\hat{\theta}) + \frac{1}{2}\sum_{\gamma\varepsilon\Gamma} N_\gamma [3f(\hat{\theta}) - 2f(\hat{\theta} - \frac{\gamma}{T}) \\ & - f(\hat{\theta} - \frac{2\gamma}{T}) - \frac{2\gamma}{T} f'(\hat{\theta}) - \frac{2\gamma}{T} f'(\hat{\theta} - \frac{\gamma}{T})] \\ & + \frac{1}{2}\sum_{\alpha\varepsilon\Gamma}\sum_{\gamma\varepsilon\Gamma} N_\alpha N_\gamma \; f(\hat{\theta} - \frac{\alpha+\gamma}{T}) \\ & - 2f(\hat{\theta} - \frac{\gamma}{T}) + f(\hat{\theta}) - \frac{2\alpha}{T} f'(\hat{\theta}) \\ & + \frac{2\alpha}{T} f'(\hat{\theta} - \frac{\gamma}{T}) + \frac{\alpha\gamma}{T^2} f''(\hat{\theta})] \quad . \end{aligned} \tag{4.2}$$

By hypothesis there exist $M > 0$, $\delta > 0$, and a Q such that $|f'''(t)| < M$ for every $t \in [\theta - \delta, \theta + \delta]$, and $|\gamma| < Q$ for every $\gamma \in \Gamma$. Since $\hat{\theta} \underset{p}{\to} \theta$ as $T \to \infty$ and $\gamma/T \to 0$ as $T \to \infty$ for every $\gamma \in \Gamma$, then for $\varepsilon > 0$ there exists $T_\varepsilon > 0$ and a family of random variables $\{r_X\}$, $|r_X| \leq 1$ a.e., such that for $T > T_\varepsilon$ the following expansions are valid with probability greater than $1 - \varepsilon$.

$$\text{(i)} \quad f(\hat{\theta} - \frac{\gamma}{T}) = f(\hat{\theta}) - \frac{\gamma}{T} f'(\hat{\theta}) + \frac{\gamma^2}{2T^2} f''(\hat{\theta} - \frac{\gamma}{T} r_\gamma)$$

$$\text{(ii)} \quad f(\hat{\theta} - \frac{2\gamma}{T}) = f(\hat{\theta}) - \frac{2\gamma}{T} f'(\hat{\theta}) + \frac{2\gamma^2}{T^2} f''(\hat{\theta} - \frac{2\gamma}{T} r_\gamma)$$

$$\text{(iii)} \quad f'(\hat{\theta} - \frac{\gamma}{T}) = f'(\hat{\theta}) - \frac{\gamma}{T} f''(\hat{\theta} - \frac{\gamma}{T} r_\gamma)$$

$$\text{(iv)} \quad f(\hat{\theta} - \frac{\alpha+\gamma}{T}) = f(\hat{\theta}) - \frac{\alpha + \gamma}{T} f'(\hat{\theta})$$

$$+ \frac{(\alpha + \gamma)^2}{2T^2} f''(\hat{\theta}) - \frac{(\alpha + \gamma)^3}{6T^3} f'''(\hat{\theta} - \frac{\alpha + \gamma}{T} r_{\alpha\gamma})$$

$$\text{(v)} \quad f(\hat{\theta} - \frac{\gamma}{T}) = f(\hat{\theta}) - \frac{\gamma}{T} f'(\hat{\theta}) + \frac{\gamma^2}{2T^2} f''(\hat{\theta})$$

$$- \frac{\gamma^3}{6T^3} f'''(\hat{\theta} - \frac{\gamma}{T} r_\gamma)$$

$$\text{(vi)} \quad f'(\hat{\theta} - \frac{\gamma}{T}) = f'(\hat{\theta}) - \frac{\gamma}{T} f''(\hat{\theta}) + \frac{\gamma^2}{2T^2} f'''(\hat{\theta} - \frac{\gamma}{T} r_\gamma)$$

Substituting (i) through (vi) into (4.2) and simplifying we obtain

$$J_\infty^{(2)}[f(\hat{\theta})] = f(\hat{\theta}) + \frac{1}{2} \sum_{\gamma\varepsilon\Gamma} \frac{\gamma^2 N_\gamma}{T^2} [- f''(\hat{\theta} - \frac{\gamma}{T} r_\gamma)$$

$$- 2f''(\hat{\theta} - \frac{2\gamma}{T} r_\gamma) + 2f''(\hat{\theta} - \frac{\gamma}{T} r_\gamma)]$$

$$+ \sum_{\alpha\varepsilon\Gamma} \sum_{\gamma\varepsilon\Gamma} N_\alpha N_\gamma [- \frac{(\alpha + \gamma)^3}{6T^3} f'''(\hat{\theta}$$

$$- \frac{\alpha + \gamma}{T} r_{\alpha\gamma}) + \frac{\gamma^3}{3T^3} f'''(\hat{\theta} - \frac{\gamma}{T} r_\gamma)$$

$$+ \frac{\gamma^2 \alpha}{T^3} f'''(\hat{\theta} - \frac{\gamma}{T} r_\gamma)] \tag{4.3}$$

with probability greater than $1 - \varepsilon$ when $T > T_\varepsilon$.

Since f'' is continuous in $[\theta - \delta, \theta + \delta]$, then there exists $M_1 > 0$ such that $|f''(t)| < M_1$ for every $t \varepsilon [\theta - \delta, \theta + \delta]$. Hence using equation (4.3) we can write

$$\left|\sqrt{T}[J_\infty^{(2)}[f(\hat{\theta})] - f(\theta)]\right|$$

$$\leq \left| \frac{\sqrt{T}}{2} \sum_{\alpha\varepsilon\Gamma} \frac{\gamma^2 N_\gamma}{T^2} [-f''(\hat{\theta} - \frac{\gamma}{T} r_\gamma) - 2f''(\hat{\theta} - \frac{2\gamma}{T} r_\gamma) \right.$$

$$\left. + 2f''(\hat{\theta} - \frac{\gamma}{T} r_\gamma)] \right|$$

$$+ \left| \sqrt{T} \sum_{\alpha\varepsilon\Gamma} \sum_{\gamma\varepsilon\Gamma} N_\alpha N_\gamma \left[\frac{-(\alpha + \gamma)^3}{6T^3} f'''(\hat{\theta} - \frac{\alpha + \gamma}{T} r_{\alpha\gamma}) \right. \right.$$

$$+ \frac{\gamma^3}{3T^3} f'''(\hat{\theta} - \frac{\gamma}{T} r_\gamma) + \frac{\gamma^2\alpha}{T^3} f'''(\hat{\theta} - \frac{\gamma}{T} r_\gamma)]$$

$$\leq \frac{5M_1R^2}{2T^{3/2}} \sum_{\gamma\varepsilon\Gamma} N_\gamma + \frac{8MR^3}{3T^{5/2}} \sum_{\alpha\varepsilon\Gamma} \sum_{\gamma\varepsilon\Gamma} N_\alpha N_\gamma$$

$$= \frac{5M_1R^2}{2} \left(\frac{N}{T^{3/2}}\right) + \frac{8MR^3}{3} \left(\frac{N}{T^{5/4}}\right)^2 \tag{4.4}$$

with probability greater than $1 - \varepsilon$ when $T > T_\varepsilon$. Since the right side of (4.4) converges to zero in probability as $T \to \infty$, then we can conclude that

$$\sqrt{T}\ [J_\infty^{(2)}[f(\hat{\theta})] - f(\hat{\theta})] \underset{p}{\to} 0$$

as $T \to \infty$.

Some discussion should be made at this point concerning the assumption that $T^{-5/4}N \underset{p}{\to} 0$ as $t \to \infty$. In order to show that this is not an unrealistic assumption, we will discuss some cases in which this condition is satisfied.

If $T^{-1}N \underset{p}{\to} C$ as $T \to \infty$, where C is a constant, then clearly $T^{-5/4}N \underset{p}{\to} 0$ as $T \to \infty$. This is the case when the stochastic process which generates N has stationary independent increments.

If with probability one, every realization of $\{I_X(t)\,|\,t \in [a,\infty)\}$ is a step function and inf $\{\gamma\} = S > 0$ or sup $\{\gamma\} = S < 0$, then

$$|\theta| = \left|\frac{1}{T} \sum_{\gamma\in\Gamma} \gamma N_\gamma\right| \geq \frac{|S|}{T} \sum_{\gamma\in\Gamma} N_\gamma = |S| \left[\frac{N}{T}\right] \geq 0 . \tag{4.5}$$

Hence from equation (4.5) we have $T^{-5/4}N \underset{p}{\to} 0$ as $T \to \infty$, since $T^{-1/4}\hat{\theta} \underset{p}{\to} 0$ as $T \to \infty$. An obvious example in this setting is when $\{I_X(t)\,|\,t \in [a,b)\}$ is a Poisson process and $\hat{\theta} = N/T$.

Combining Theorem 4.1 and equation IV.4.7 we conclude our development with the following theorem.

<u>THEOREM 4.2.</u> Let the conditions of Theorem 4.1 be satisfied. If $\hat{\sigma}_T^2 \underset{p}{\to} \sigma^2$ as $T \to \infty$, then

$$\frac{J_\infty^{(2)}[f(\hat{\theta})] - f(\theta)}{\sqrt{\sum_{\gamma\in\Gamma} N_\gamma [f(\hat{\theta} - \frac{\gamma}{T}) - f(\hat{\theta})]^2}} \underset{\mathcal{L}}{\to} N(0,1)$$

as $T \to \infty$.

APPENDIX

A PROOF THAT $J^{(2)}[f(\hat{\theta});n] \to J_{\infty}^{(2)}[f(\hat{\theta})]$ a.e. AS $n \to \infty$

BY J. E. ADAMS

For ease in notation and readability, some preliminary discussion is necessary before the proof of the convergence theorem is presented.

When f is a function with a continuous second derivative in an interval containing x, we will adopt the following notation

$$A_1(n) = \frac{f[x + \alpha(n) + h(n)] - f[x + \alpha(n)]}{h(n)}$$

and

$$A_2(n) = \frac{f[x + h(n) + k(n)] - f[x + h(n)]}{h(n)\ k(n)}$$

$$+ \frac{-f[x + k(n)] + f(x)}{h(n)\ k(n)}$$

where $h(n) \to 0$, $k(n) \to 0$, $\alpha(n) \to a$ as $n \to \infty$. It is easy to show that

$$\lim_{n\to\infty} A_1(n) = f'(x + a)$$

and

$$\lim_{n\to\infty} A_2(n) = f''(x) .$$

In many cases the forms defined above will depend on the elements of some index set. The notations $A_1(n,i)$, $A_1(n,i,j)$, $i \neq j$, $A_2(n,i)$, $A_2(n,i,j)$, $i \neq j$, will be adopted to indicate the dependency of these forms on i and j.

The definitions of Γ, N_γ, and N have already been stated in the definition of $J_\infty[f(\hat{\theta})]$; however at this point, we need to be more specific in their structure.

Suppose the stochastic process, $\{I_X(t) | t \in [a,b]\}$, which is completely determined by the stochastic process $\{X(t) | t \in [a,b]\}$, is piecewise continuous on [a,b]. Let the set $\mathscr{S}$ be defined by

$$\mathscr{S} = \{x | x \text{ is a realization of } \{X(t) | t \in [a,b]\}\}$$

and let

$$\mathscr{S}_1 = \{x | x \in \mathscr{S} \text{ and } I_x \text{ is piecewise continuous on } [a,b]\}$$

where I_x is a realization of $\{I_X(t) | t \in [a,b]\}$ determined by x. Then for each $x \in \mathscr{S}_1$ we define

$$H_x(a) = I_x(a^+) - I_x(a)$$

$$H_x(b) = I_x(b) - I_x(b^-)$$

and

$$H_x(t) = I_x(t^+) - I_x(t^-)$$

for each $t \in (a,b)$. Also for $x \notin \mathscr{S}_1$ we define

$$H_x(t) = 0$$

for each $t \in [a,b]$. It was shown in [48] that for each $t \in [a,b]$, $H_x(t)$ is a random variable.

Now let R_t denote the range of $H_x(t)$ and let Γ be defined by

$$\Gamma = \bigcup_{t\varepsilon[a,b]} R_t - \{0\} \quad .$$

Also, for each $\alpha \varepsilon \Gamma$, let N_α be defined on $\mathscr{S}$ by

$$N_\alpha(x) = \mathcal{C}[t \mid t \varepsilon [a,b], x \varepsilon \mathscr{S}, H_x(t) = \alpha]$$

where $\mathcal{C}(\cdot)$ denotes the cardinality of a set. Now define

$$N(x) = \sum_{\alpha\varepsilon\Gamma} N_\alpha(x) \quad .$$

As was stated in the definition of $J_\infty[f(\hat{\theta})]$, $N_\alpha(x)$

is simply the number of points in [a,b] where I_x has a jump of size α, and N(x) is the total number of points in [a,b] where I_x has a jump. Also note that if $x \in \mathcal{S}_1$, N(x) is finite and only a finite number of the $N_\alpha(x)$ are nonzero, and if $x \notin \mathcal{S} - \mathcal{S}_1$, then N(x) = 0.

<u>THEOREM</u>. Let $J^{(2)}[f(\hat{\theta});n]$ and $J_\infty^{(2)}[f(\hat{\theta})]$ be as defined in Definitions 2.1 and 2.2, respectively, of Chapter V. Suppose that the second derivative of f is continuous over the range of $\hat{\theta}$. If the stochastic process $\{I_X(t) | t \in [a,b]\}$ is of bounded variation, and if for each $t \in [a,b]$, $P[H_X(t) = 0] = 1$, where $H_X(t)$ is defined above, then $J^{(2)}[f(\hat{\theta});n] \to J_\infty^{(2)}[f(\hat{\theta})]$ a.e. as $n \to \infty$.

Proof: Much of the following preliminary discussion can be found in [48], however, it is included here for completeness and convenience to the reader.

Let

$$A = \bigcup_{n=1}^{\infty} \{a + \frac{m}{n}(b - a) | m \text{ is an integer and } 0 \le m \le n\} .$$

Then, since A is countable,

$$P[H_X(t) = 0 \text{ for every } t \in A] = 1 . \tag{1}$$

Let $\mathcal{J}$ denote the set of all realizations of

$\{X(t)|t \in S\}$ such that for each $x \in \mathcal{J}$

(i) I_x is piecewise continuous on [a,b],

(ii) I_x is of bounded variation on [a,b],

(iii) $H_x(t) = 0$ for every $t \in A$,

where I_x is the realization of $\{I_X(t)|t \in [a,b]\}$ determined by x. Now since $\{I_X(t)|t \in [a,b]\}$ is of bounded variation on [a,b] and piecewise continuous, and equation (1) holds, then $P\{x$ such that $x \in \mathcal{J}\} = 1$. Thus the proof of the theorem is complete if for $x \in \mathcal{J}$ $\lim_{n\to\infty} J^{(2)}[f(\hat{\theta});n]\Big|_x = J_\infty^{(2)}[f(\hat{\theta})]\Big|_x$.

Now let $x \in \mathcal{J}$ and let S_x and C_x be defined for $t \in [a,b]$ by

$$S_x(t) = \sum_{\omega(t)} H_x(u) \quad , \tag{2}$$

where $\omega(t) = \{u:a \leq u \leq t \text{ and } H_x(u) \neq 0\}$, and

$$C_x(t) = I_x(t^+) - S_x(t) \quad .$$

Since H_x is nonzero at only a finite number of points of [a,b], then S_x is finite and a step function. Also note that $C_x(t)$ is continuous and therefore uniformly continuous on [a,b]. Further, for $u,v \in A$,

$$I_x(u) - I_x(v)$$

$$= C_x(u) + S_x(u) - S_x(v) - C_x(v) \quad . \tag{3}$$

Let $\Delta_n = \{t_0, t_1, \ldots, t_n\}$ denote a partition of $[a,b]$ where $a = t_0 < t_1 < \ldots < t_n = b$, $T = b - a$ and $t_i - t_{i-1} = T/n$. Then for $\varepsilon > 0$ there exists n_ε such that

$$\left| C_x(t_i) - C_x(t_{i-1}) \right| < \varepsilon$$

$$\mathcal{C}[\{t \mid t \in [t_{i-1}, t_i], H_x(t) \neq 0\}] \leq 1 \tag{4}$$

when $n > n_\varepsilon$ and $t_i, t_{i-1} \in \Delta_n$.

Also define

$$I_0 = \{i \mid H_x(t) = 0 \text{ for every } t \in [t_{i-1}, t_i],$$
$$t_{i-1}, t_i \in \Delta_n\}$$
$$(5)$$
$$I_\alpha = \{i \mid H_x(t) = \alpha \text{ for some } t \in [t_{i-1}, t_i],$$
$$t_i, t_{i-1} \in \Delta_n\} \quad .$$

Although not explicit in the notation, note that I_0 and I_α, $\alpha \varepsilon \Gamma$ are dependent on n.

Then for $n > n_\varepsilon$

$$I_0 \cap I_\alpha = \phi \text{ for each } \alpha \varepsilon \Gamma \text{ and}$$

$$I_0 \cup [\bigcup_{\alpha\varepsilon\Gamma} I_\alpha] = \{1,2,\ldots,n\} \text{ and}$$

$$I_0^c = \bigcup_{\alpha\varepsilon\Gamma} I_\alpha \triangleq I \quad .$$

In the remainder of this proof we will in many instances use the same notation for the estimate as for the estimator. Also n will be sufficiently large for (4) to hold, and for convenience in notation we write $N_\alpha(x) = N_\alpha$ and $N(x) = N$.

Lemma 1.

$$\sum_{i\varepsilon I_0} \frac{\hat{\theta}_i}{n} \to \left[\hat{\theta} - \sum_{\alpha\varepsilon\Gamma} \frac{\alpha N_\alpha}{T} \right]$$

as $n \to \infty$ where $\hat{\theta}_i$ is defined in Definition 2.4 of Chapter IV.

Proof: Let $\varepsilon > 0$, then there exists $n'_\varepsilon > 0$ such that, for $n \geq n'_\varepsilon$ and $t_i, t_{i-1} \varepsilon \Delta_n$,

$$\left| C_x(t_i) - C_x(t_{i-1}) \right| < \frac{\varepsilon T}{N + 1} \quad .$$

Hence

$$\sum_{\alpha\varepsilon\Gamma} \sum_{i\varepsilon I_\alpha} \left| \frac{C_x(t_i) - C_x(t_{i-1})}{T} \right| < \frac{\varepsilon T \sum_{\alpha\varepsilon\Gamma} N_\alpha}{(N + 1)T} < \varepsilon. \quad (6)$$

Thus with $\hat{\theta} - \sum_{\alpha\varepsilon\Gamma} \frac{\alpha N_\alpha}{T} = A$ we have

$$\left| \sum_{i\varepsilon I_0} \frac{\hat{\theta}_i}{n} - A \right|$$

$$= \left| \sum_{i\varepsilon I_0} \frac{I_x(t_i) - I_x(t_{i-1})}{T} - A \right|$$

$$= \left| \sum_{i\varepsilon I_0} \frac{\Delta C_x(t_i)}{T} - A \right| ,$$

by (3), where

$$\Delta C_x(t_i) = C_x(t_i) - C_x(t_{i-1})$$

$$= \left| \sum_{i=1}^{n} \frac{\Delta C_x(t_i)}{T} - A - \sum_{\alpha\varepsilon\Gamma} \sum_{i\varepsilon I_\alpha} \frac{\Delta C_x(t_i)}{T} \right|$$

$$\leq \left| \frac{C_x(b) - C_x(a)}{T} - A \right| + \sum_{\alpha\varepsilon\Gamma} \sum_{i\varepsilon I_\alpha} \left| \frac{\Delta C_x(t_i)}{T} \right|$$

$$< \varepsilon$$

since $\dfrac{C_x(b) - C_x(a)}{T} = \hat{\theta} - \sum_{\alpha \varepsilon \Gamma} \dfrac{\alpha N_\alpha}{T}$ and by equation (6).

<u>Lemma 2.</u>

$$\sum_{i \varepsilon I} \frac{\hat{\theta}_i}{n} \to \sum_{\alpha \varepsilon \Gamma} \frac{\alpha N_\alpha}{T}$$

as $n \to \infty$.

Proof: Since $\hat{\theta}$ is independent of n and

$$\sum_{i \varepsilon I} \frac{\hat{\theta}_i}{n} = \sum_{\alpha \varepsilon \Gamma} \sum_{i \varepsilon I_\alpha} \frac{\hat{\theta}_i}{n} = \hat{\theta} - \sum_{i \varepsilon I_0} \frac{\hat{\theta}_i}{n}$$

then by Lemma 1 we have

$$\lim_{n \to \infty} \sum_{i \varepsilon I} \frac{\hat{\theta}_i}{n} = \hat{\theta} - \left(\hat{\theta} - \sum_{\alpha \varepsilon \Gamma} \frac{\alpha N_\alpha}{T} \right) = \sum_{\alpha \varepsilon \Gamma} \frac{\alpha N_\alpha}{T} .$$

<u>Lemma 3.</u>

$$\sum_{\substack{I_0, I_0 \\ i \neq j}} \frac{\hat{\theta}_i \hat{\theta}_j}{n^2} \to \left[\hat{\theta} - \sum_{\alpha \varepsilon \Gamma} \frac{\alpha N_\alpha}{T} \right]^2$$

as $n \to \infty$.

Proof:

$$\sum_{\substack{I_0, I_0 \\ i \neq j}} \frac{\hat{\theta}_i \hat{\theta}_j}{n^2} = \left(\sum_{i \varepsilon I_0} \frac{\hat{\theta}_i}{n} \right)^2 - \sum_{i \varepsilon I_0} \frac{\hat{\theta}_i^{\,2}}{n^2} .$$

By Lemma 1

$$\left(\sum_{i \varepsilon I_0} \frac{\hat{\theta}_i}{n} \right)^2 \to \left(\hat{\theta} - \sum_{\alpha \varepsilon \Gamma} \frac{\alpha N_\alpha}{T} \right)^2$$

as $n \to \infty$. Now consider

$$\left| \sum_{i \varepsilon I_0} \frac{\hat{\theta}_i^{\,2}}{n^2} \right| = \left| \sum_{i \varepsilon I_0} \left[\frac{\Delta I_x(t_i)}{T} \right]^2 \right|$$

$$\leq \left| \max_{i \varepsilon I_0} \left\{ \frac{\Delta I_x(t_i)}{T} \right\} \right|$$

$$\cdot \sum_{i \varepsilon I_0} \frac{\left| \Delta I_x(t_i) \right|}{T}$$

$$\leq \left| \max_{i \varepsilon I_0} \left\{ \frac{\Delta C_x(t_i)}{} \right\} \right| V_{[a,b]}(I_x)$$

where $V_{[a,b]}(I_x)$ denotes the total variation of I_x on [a,b]. Now by (4) the first factor above can be made arbitrarily small by choosing n sufficiently large. Hence

$$\sum_{i \varepsilon I_0} \frac{\hat{\theta}_i^2}{n^2} \to 0$$

as $n \to \infty$. Therefore

$$\sum_{\substack{I_0, I_0 \\ i \neq j}} \frac{\hat{\theta}_i \hat{\theta}_j}{n^2} \to \left[\hat{\theta} - \sum_{\alpha \varepsilon \Gamma} \frac{\alpha N_\alpha}{T} \right]^2$$

as $n \to \infty$.

<u>Lemma 4</u>. Given that I_x is of bounded variation on [a,b] then

$$\lim_{n \to \infty} \sum_{\substack{I_0, I_0 \\ i \neq j}} \frac{\left| \hat{\theta}_i \hat{\theta}_j \right|}{n^2} \leq \left[V_{[a,b]}(I_x) \right]^2 .$$

Proof: Clearly $\lim_{n\to\infty} \sum_{i\varepsilon I_0} \frac{|\hat{\theta}_i|}{n}$ exists since

$$\sum_{i\varepsilon I_0} \frac{|\hat{\theta}_i|}{n} \leq V_{[a,b]}(I_x)$$ for every n. Now follow

the steps of Lemma 3 with absolute values from the beginning.

Lemma 5. For $i \varepsilon I_\alpha$, $\hat{\theta}_i/n \to \alpha/T$ as $n \to \infty$.

Proof:

$$\left|\frac{\hat{\theta}_i}{n} - \frac{\alpha}{T}\right| = \left|\frac{\Delta I_x(t_i)}{T} - \frac{\alpha}{T}\right|$$

$$= \left|\frac{\Delta S_x(t_i)}{T} - \frac{\alpha}{T} + \frac{\Delta C_x(t_i)}{T}\right|$$

$$= \left|\frac{\Delta C_x(t_i)}{T}\right| < \varepsilon/T$$

for n sufficiently large.

By appropriate algebraic manipulations, one can show that

$$J^{(2)}[f(\hat{\theta});n] = \frac{n}{2(n-1)} \sum_{i \neq j} [f(\hat{\theta}) - 2f(\hat{\theta}^{i}) + f(\hat{\theta}^{ij})] + \frac{2}{n} \sum_{i \neq j} [f(\hat{\theta}^{i}) - f(\hat{\theta}^{ij})] + \frac{1}{n} \sum_{i=1} f(\hat{\theta}^{i}) \; . \qquad (7)$$

We now define

$$\phi(\hat{\theta},\hat{\theta}^{i},\hat{\theta}^{ij}) = f(\hat{\theta}) - 2f(\hat{\theta}^{i}) + f(\hat{\theta}^{ij})$$

$$\phi(\hat{\theta}^{i},\hat{\theta}^{ij}) = f(\hat{\theta}^{i}) - f(\hat{\theta}^{ij})$$

$$\phi(\hat{\theta},\hat{\theta}^{i}) = f(\hat{\theta}) - f(\hat{\theta}^{i}) \; .$$

Now with I_0 defined by equation (5) and $I = I_0^c$ using equation (7) we can write $J^{(2)}[f(\hat{\theta});n]$ in the following form, where the sums are over the indicated sets.

$$J^{(2)}[f(\hat{\theta});n] = \frac{n}{2(n-1)} \sum_{\substack{I_0,I_0 \\ i \neq j}} \phi(\hat{\theta},\hat{\theta}^{i},\hat{\theta}^{ij})$$

$$+ \frac{2}{n} \sum_{\substack{I_0, I_0 \\ i \neq j}} \phi(\hat{\theta}^i, \hat{\theta}^{ij}) + \frac{n}{2(n-1)} \sum_{\substack{I, I \\ i \neq j}} \phi(\hat{\theta}, \hat{\theta}^i, \hat{\theta}^{ij})$$

$$+ \frac{2}{n} \sum_{\substack{I, I \\ i \neq j}} \phi(\hat{\theta}^i, \hat{\theta}^{ij}) + \frac{n}{2(n-1)} \sum_{\substack{I_0, I \\ i \neq j}} \phi(\hat{\theta}, \hat{\theta}^i)$$

$$- \frac{n}{n-1} \sum_{\substack{I, I_0 \\ i \neq j}} \phi(\hat{\theta}^i, \hat{\theta}^{ij})$$

$$+ \frac{n}{2(n-1)} \left[\sum_{\substack{I, I_0 \\ i \neq j}} \phi(\hat{\theta}, \hat{\theta}^{ij}) - \sum_{\substack{I_0, I \\ i \neq j}} \phi(\hat{\theta}^i, \hat{\theta}^{ij}) \right]$$

$$+ \frac{2}{n} \sum_{\substack{I_0, I \\ i \neq j}} \phi(\hat{\theta}^i, \hat{\theta}^{ij}) + \frac{2}{n} \sum_{\substack{I, I_0 \\ i \neq j}} \phi(\hat{\theta}^i, \hat{\theta}^{ij})$$

$$+ \frac{1}{n} \sum_{i=1}^{n} f(\hat{\theta}^i) \quad . \tag{8}$$

The following identity is given for the first term in (8) for use in the analysis which follows.

$$\frac{n}{2(n-1)} \sum_{\substack{I_0, I_0 \\ i \neq j}} [f(\hat{\theta}) - 2f(\hat{\theta}^i) + f(\hat{\theta}^{ij})]$$

$$= \frac{n}{2(n-1)} \sum_{\substack{I_0, I_0 \\ i \neq j}} \left[f(\hat{\theta}) - 2f\left(\hat{\theta} - \frac{\hat{\theta}_i - \hat{\theta}}{n-1}\right) \right.$$

$$\left. + f\left(\hat{\theta} - \frac{\hat{\theta}_i + \hat{\theta}_j - 2\hat{\theta}}{n-2}\right)\right]$$

$$= \frac{n}{2(n-1)} \sum_{\substack{I_0, I_0 \\ i \neq j}} \left[\left[f\left(\hat{\theta} - \frac{\hat{\theta}_i - \hat{\theta}}{n-2} - \frac{\hat{\theta}_j - \hat{\theta}}{n-2}\right) \right. \right.$$

$$- f\left(\hat{\theta} - \frac{\hat{\theta}_j - \hat{\theta}}{n-2}\right) - f\left(\hat{\theta} - \frac{\hat{\theta}_i - \hat{\theta}}{n-2}\right)$$

$$\left. \left. + f(\hat{\theta}) \right] \Big/ \frac{(\hat{\theta}_i - \hat{\theta})(\hat{\theta}_j - \hat{\theta})}{(n-2)^2} \right] \frac{(\hat{\theta}_i - \hat{\theta})(\hat{\theta}_j - \hat{\theta})}{(n-2)^2}$$

$$+ \frac{n}{2(n-1)} \sum_{\substack{I_0, I_0 \\ i \neq j}} \left[\left[f\left(\hat{\theta} - \frac{\hat{\theta}_i - \hat{\theta}}{n-2}\right) - f\left(\hat{\theta} - \frac{\hat{\theta}_i - \hat{\theta}}{n-2}\right. \right. \right.$$

$$+ \frac{\hat{\theta}_i - \hat{\theta}}{(n-2)(n-1)}\Bigg)\Bigg] \Bigg/ \frac{(\hat{\theta} - \hat{\theta}_i)}{(n-1)(n-2)}\Bigg]$$

$$\cdot \frac{(\hat{\theta} - \hat{\theta}_i)}{(n-1)(n-2)} + \frac{n}{2(n-1)} \sum_{\substack{I_0, I_0 \\ i \neq j}} \Bigg[\Bigg[f\Bigg(\hat{\theta} - \frac{\hat{\theta}_j - \hat{\theta}}{n-1}$$

$$- \frac{\hat{\theta}_j - \hat{\theta}}{(n-1)(n-2)}\Bigg) - f\Bigg(\hat{\theta} - \frac{\hat{\theta}_j - \hat{\theta}}{n-1}\Bigg)$$

$$- f\Bigg(\hat{\theta} - \frac{\hat{\theta}_j - \hat{\theta}}{(n-1)(n-2)}\Bigg)$$

$$+ f(\hat{\theta})\Bigg] \Bigg/ \Bigg(\frac{\hat{\theta}_j - \hat{\theta}}{n-1}\Bigg)\Bigg(\frac{\hat{\theta}_j - \hat{\theta}}{(n-1)(n-2)}\Bigg)\Bigg]$$

$$\cdot \frac{(\hat{\theta}_j - \hat{\theta})^2}{(n-1)^2(n-2)} + \frac{n}{2(n-1)} \sum_{\substack{I_0, I_0 \\ i \neq j}} \Bigg[f\Bigg(\hat{\theta} - \frac{\hat{\theta}_j - \hat{\theta}}{n-1}\Bigg)$$

$$- f\left(\hat{\theta} - \frac{\hat{\theta}_i - \hat{\theta}}{n - 1}\right)\Bigg]$$

$$+ \frac{n}{2(n - 1)} \sum_{\substack{I_0, I_0 \\ i \neq j}} \left[\left[f\left(\hat{\theta} - \frac{\hat{\theta}_j - \hat{\theta}}{(n - 1)(n - 2)}\right)\right.\right.$$

$$\left. - f(\hat{\theta})\right] \Bigg/ \left.\frac{\hat{\theta}_j - \hat{\theta}}{(n - 1)(n - 2)}\right]$$

$$\cdot \frac{\hat{\theta}_j - \hat{\theta}}{(n - 1)(n - 2)} \qquad .$$

We will now consider the first term in (8). Since $\lim_{n \to \infty} n/(n - 1) = 1$, the coefficient $n/(n - 1)$ will not affect the convergence of the terms in equation (8). We therefore have

$$\left| \sum_{\substack{I_0, I_0 \\ i \neq j}} \phi(\hat{\theta}, \hat{\theta}^i, \hat{\theta}^{ij}) - f''(\hat{\theta}) \left[\sum_{\alpha \varepsilon \Gamma} \frac{\alpha N_\alpha}{T}\right]^2 \right.$$

$$- 2f'(\hat{\theta}) \left[\sum_{\alpha\varepsilon\Gamma} \frac{\alpha N_\alpha}{T} \right] \Bigg|$$

$$= \Bigg| \sum_{\substack{I_0, I_0 \\ i \neq j}} A_2(n,i,j) \frac{(\hat{\theta} - \hat{\theta}_i)(\hat{\theta} - \hat{\theta}_j)}{(n - 2)^2}$$

$$- f''(\hat{\theta}) \left[\sum_{\alpha\varepsilon\Gamma} \frac{\alpha N_\alpha}{T} \right]^2 + \sum_{\substack{I_0, I_0 \\ i \neq j}} A_1(n,i) \frac{(\hat{\theta}_i - \hat{\theta})}{(n - 1)(n - 2)}$$

$$- f'(\hat{\theta}) \sum_{\alpha\varepsilon\Gamma} \frac{\alpha N_\alpha}{T} + \sum_{\substack{I_0, I_0 \\ i \neq j}} A_2(n,j) \frac{(\hat{\theta} - \hat{\theta}_j)^2}{(n - 1)^2 (n - 2)}$$

$$- \sum_{\substack{I_0, I_0 \\ i \neq j}} f''(\hat{\theta}) \frac{(\hat{\theta} - \hat{\theta}_j)^2}{(n - 1)^2 (n - 2)}$$

$$+ \sum_{\substack{I_0, I_0 \\ i \neq j}} f''(\hat{\theta}) \frac{(\hat{\theta} - \hat{\theta}_j)^2}{(n - 1)^2 (n - 2)} + \sum_{\substack{I_0, I_0 \\ i \neq j}} [f(\hat{\theta}^i) - f(\hat{\theta}^j)]$$

$$+ \sum_{\substack{I_0, I_0 \\ i \neq j}} A_1(n,j) \frac{(\hat{\theta} - \hat{\theta}_j)}{(n-1)(n-2)} - f'(\hat{\theta}) \sum_{\alpha \varepsilon \Gamma} \frac{\alpha N_\alpha}{T} \Bigg|$$

$$\leq \Bigg| \sum_{\substack{I_0, I_0 \\ i \neq j}} A_2(n,i,j) \frac{(\hat{\theta} - \hat{\theta}_i)(\hat{\theta} - \hat{\theta}_j)}{(n-2)^2}$$

$$- f''(\hat{\theta}) \left[\sum_{\alpha \varepsilon \Gamma} \frac{\alpha N_\alpha}{T} \right]^2 \Bigg| + \Bigg| \sum_{\substack{I_0, I_0 \\ i \neq j}} A_1(n,i) \frac{(\hat{\theta} - \hat{\theta}_i)}{(n-1)(n-2)}$$

$$- f'(\hat{\theta}) \sum_{\alpha \varepsilon \Gamma} \frac{\alpha N_\alpha}{T} \Bigg|$$

$$+ \Bigg| \sum_{\substack{I_0, I_0 \\ i \neq j}} A_2(n,i) \frac{(\hat{\theta} - \hat{\theta}_j)^2}{(n-1)^2(n-2)}$$

$$- \sum_{\substack{I_0, I_0 \\ i \neq j}} f''(\hat{\theta}) \frac{(\hat{\theta} - \hat{\theta}_j)^2}{(n-1)^2(n-2)} \Bigg|$$

$$+ \frac{|f''(\hat{\theta})|}{(n-2)} \left| \sum_{\substack{I_0, I_0 \\ i \neq j}} \frac{(\hat{\theta}_j - \hat{\theta})^2}{(n-1)^2 (n-2)} \right|$$

$$+ \left| \sum_{\substack{I_0, I_0 \\ i \neq j}} A_1(n,j) \frac{(\hat{\theta}_j - \hat{\theta})}{(n-1)(n-2)} \right.$$

$$\left. - f'(\hat{\theta}) \sum_{\alpha \varepsilon \Gamma} \frac{\alpha N_\alpha}{T} \right| \quad . \qquad (9)$$

We now consider (9) term by term.

$$\left| \sum_{\substack{I_0, I_0 \\ i \neq j}} A_2(n,i,j) \frac{(\hat{\theta}_i - \hat{\theta})(\hat{\theta}_j - \hat{\theta})}{(n-2)^2} - f''(\hat{\theta}) \left[\sum_{\alpha \varepsilon \Gamma} \frac{\alpha N_\alpha}{T} \right]^2 \right|$$

$$\leq \left| \sum_{\substack{I_0, I_0 \\ i \neq j}} A_2(n,i,j) \frac{(\hat{\theta}_i - \hat{\theta})(\hat{\theta}_j - \hat{\theta})}{(n-2)^2} \right.$$

$$- \sum_{\substack{I_0, I_0 \\ i \neq j}} f''(\hat{\theta}) \frac{(\hat{\theta}_i - \hat{\theta})(\hat{\theta}_j - \hat{\theta})}{(n-2)^2} \Bigg|$$

$$+ \left| f''(\hat{\theta}) \left[\sum_{\alpha \varepsilon \Gamma} \frac{\alpha N_\alpha}{T} \right]^2 - f''(\hat{\theta}) \sum_{\substack{I_0, I_0 \\ i \neq j}} \frac{(\hat{\theta}_i - \hat{\theta})(\hat{\theta}_j - \hat{\theta})}{(n-2)^2} \right|$$

$$\leq \left| \sum_{\substack{I_0, I_0 \\ i \neq j}} \left| f''(\hat{\theta}) - A_2(n,i,j) \right| \left| \frac{(\hat{\theta} - \hat{\theta}_i)(\hat{\theta} - \hat{\theta}_j)}{(n-2)^2} \right| \right.$$

$$+ \left| f''(\hat{\theta}) \right| \left| \left[\sum_{\alpha \varepsilon \Gamma} \frac{\alpha N_\alpha}{T} \right]^2 - \sum_{\substack{I_0, I_0 \\ i \neq j}} \frac{(\hat{\theta} - \hat{\theta}_i)(\hat{\theta} - \hat{\theta}_j)}{(n-2)^2} \right| . \tag{10}$$

Now since $\hat{\theta}$ is independent of n, and fixed, we have $\hat{\theta}/n \to 0$ as $n \to \infty$. Also from Lemmas 1 and 3 it is necessary that for $i, j \in I_0$, $\hat{\theta}_i/n \to 0$ and $\hat{\theta}_i \hat{\theta}_j/n^2 \to 0$ as $n \to \infty$ and hence

$$\frac{\hat{\theta} - \hat{\theta}_i}{n} \to 0 \quad \text{and} \quad \frac{(\hat{\theta} - \hat{\theta}_i)(\hat{\theta} - \hat{\theta}_j)}{n^2} \to 0 \tag{11}$$

as $n \to \infty$. Let $\varepsilon > 0$, then for n sufficiently large

$$\left.\begin{aligned} &\left| f''(\hat{\theta}) - A_2(n,i,j) \right| < \varepsilon \\ &\left| f'(\hat{\theta}) - A_1(n,i) \right| < \varepsilon \\ &\left| f''(\hat{\theta}) - A_2(n,i) \right| < \varepsilon \\ &\left| f'(\hat{\theta}) - A_1(n,j) \right| < \varepsilon \end{aligned}\right\} \qquad (12)$$

Also for n sufficiently large, Lemmas 1 and 3 imply

$$\left| \left[\sum_{\alpha \varepsilon \Gamma} \frac{\alpha N_\alpha}{T} \right]^2 - \sum_{\substack{I_0, I_0 \\ i \neq j}} \frac{(\hat{\theta} - \hat{\theta}_i)(\hat{\theta} - \hat{\theta}_j)}{(n - 2)^2} \right| < \varepsilon .$$

Thus continuing the inequality (10) we have

$$< \varepsilon \left\{ \sum_{\substack{I_0, I_0 \\ i \neq j}} \left| \frac{(\hat{\theta} - \hat{\theta}_i)(\hat{\theta} - \hat{\theta}_j)}{(n - 2)^2} \right| + \left| f''(\hat{\theta}) \right| \right\}$$

$$\leq \varepsilon \left\{ [V_{[a,b]}(I_x)]^2 + f''(\hat{\theta}) \right\}$$

which can be made arbitrarily small by choosing ε sufficiently small and n sufficiently large.

We now examine the second term of (9).

$$\left| \frac{n - N - 1}{n - 2} \sum_{i \varepsilon I_0} A_1(n,i) \frac{(\hat{\theta} - \hat{\theta}_i)}{n - 1} - f'(\hat{\theta}) \sum_{\alpha \varepsilon \Gamma} \frac{\alpha N_\alpha}{T} \right|$$

$$\leq \left| \frac{n - N - 1}{n - 2} \sum_{i \varepsilon I_0} A_1(n,i) \frac{(\hat{\theta} - \hat{\theta}_i)}{n - 1} \right.$$

$$\left. - \frac{n - N - 1}{n} \sum_{i \varepsilon I_0} f'(\hat{\theta}) \frac{(\hat{\theta} - \hat{\theta}_i)}{n - 1} \right|$$

$$+ \left| \frac{n - N - 1}{n} \sum_{i \varepsilon I_0} f'(\hat{\theta}) \frac{(\hat{\theta} - \hat{\theta}_i)}{n - 1} \right.$$

$$\left. - f'(\hat{\theta}) \sum_{\alpha \varepsilon \Gamma} \frac{\alpha N_\alpha}{T} \right| \qquad (13)$$

$$\leq \left| \frac{n - N - 1}{n - 2} \right| \sum_{i \varepsilon I_0} \left| A_1(n,i) - f'(\hat{\theta}) \right| \frac{\left| \hat{\theta} - \hat{\theta}_i \right|}{n - 1}$$

$$+ \left| f'(\hat{\theta}) \right| \left| \frac{n - N - 1}{n} \sum_{i \varepsilon I_0} \frac{(\hat{\theta} - \hat{\theta}_i)}{n - 1} \right.$$

$$- \sum_{\alpha\varepsilon\Gamma} \frac{\alpha N_\alpha}{T} \Bigg|$$

$$\leq \varepsilon \left| \frac{n - N - 1}{n - 2} \right| \sum_{i\varepsilon I_0} \frac{|\hat{\theta} - \hat{\theta}_i|}{n - 1}$$

$$+ |f'(\hat{\theta})| \Bigg| \frac{n - N - 1}{n} \sum_{i\varepsilon I_0} \frac{(\hat{\theta} - \hat{\theta}_i)}{n - 1}$$

$$- \sum_{\alpha\varepsilon\Gamma} \frac{\alpha N_\alpha}{T} \Bigg|$$

by (13). Since $(n - N - 1)/(n - 2) \to 1$ as $n \to \infty$ and I_x is of bounded variation and by Lemma 1, we have, continuing the above inequality,

$$< \varepsilon \{V_{[a,b]}(I_x) + |f'(\hat{\theta})|\} \quad ,$$

which can be made arbitrarily small by appropriately choosing ε and n.

Note that the third term of (9) can be written

$$\frac{1}{n - 2} \Bigg| \sum_{\substack{I_0, I_0 \\ i \neq j}} A_2(n,j) \frac{(\hat{\theta} - \hat{\theta}_j)^2}{(n - 1)^2}$$

$$- \sum_{\substack{I_0, I_0 \\ i \neq j}} f''(\hat{\theta}) \frac{(\hat{\theta} - \hat{\theta}_j)^2}{(n - 1)^2} \Bigg|$$

$$= \frac{n - N - 1}{n - 2} \Bigg| \sum_{j \varepsilon I_0} A_2(n, j) \frac{(\hat{\theta} - \hat{\theta}_j)^2}{(n - 1)^2}$$

$$- \sum_{j \varepsilon I_0} f''(\hat{\theta}) \frac{(\hat{\theta} - \hat{\theta}_j)^2}{(n - 1)^2} \Bigg| \quad . \qquad (14)$$

Now arguing the same as for the first term of (9), it is easy to show that (14) can be made arbitrarily small by choosing n sufficiently large.

The fourth term of (9) is

$$\frac{(n - N - 1)\ f''(\hat{\theta})}{(n - 2)^2} \Bigg| \sum_{j \varepsilon I_0} \frac{(\hat{\theta}_j - \hat{\theta})^2}{(n - 1)^2} \Bigg|$$

$$\leq \frac{|f''(\hat{\theta})|}{(n - 2)} \sum_{j \varepsilon I_0} \frac{|\hat{\theta}_j - \hat{\theta}|^2}{(n - 1)^2}$$

and

$$\lim_{n\to\infty} \sum_{j\varepsilon I_0} \frac{\left|\hat{\theta}_j - \hat{\theta}\right|^2}{(n - 1)^2}$$

exists since I_x is of bounded variation. Then clearly

$$\lim_{n\to\infty} \frac{|f''(\hat{\theta})|}{(n - 2)} \left| \sum_{\substack{I_0, I_0 \\ i\neq j}} \frac{(\hat{\theta}_j - \hat{\theta})^2}{(n - 1)^2 (n - 2)} \right| = 0 . \quad (15)$$

The argument to show the fifth term of (9) can be made arbitrarily small with n sufficiently large is identical to the argument used for the second term of (9).

Therefore, combining the above results we have

$$\lim_{n\to\infty} \frac{n}{2(n - 1)} \sum_{\substack{I_0, I_0 \\ i\neq j}} \phi(\hat{\theta}, \hat{\theta}^i, \hat{\theta}^j)$$

$$= \frac{1}{2} f''(\hat{\theta}) \left[\sum_{\alpha\varepsilon\Gamma} \frac{\alpha N_\alpha}{T} \right]^2 + f'(\hat{\theta}) \sum_{\alpha\varepsilon\Gamma} \frac{\alpha N_\alpha}{T} . \quad (16)$$

A similar analysis can be given for each of the remaining terms. But as is clear from above the algebra is rather tiring. However, completion of such analysis will yield the remaining terms of $J_{\infty}^{(2)}[f(\hat{\theta})]$ and hence complete the proof. Since there are really no new thoughts in producing these additional terms they will not be given here and hence the theorem is established. The reader who is interested in a detailed analysis of the terms not considered above can find them in [1].

REFERENCES

[1]. Adams, J. E. [1971], Jackknifing stochastic process using the second order J_∞-estimator," Ph.D. dissertation, Texas Tech University.

[2]. Adams, J. E., Gray, H. L., and Watkins, T. A. [1971], "An asymptotic characterization of bias reduction by jackknifing," Ann. Math. Statist. 42, No. 5.

[3]. Arvesen, J. [1969], "Jackknifing U-statistics," Ann. Math. Statist. 40, 2076-2100.

[4]. Arvesen, J., and Schmitz, T. [1970], "Robust procedures for variance component problems using the jackknife," Biometrics 26, 677-686.

[5]. Bartlett, M. S. [1952], "An inverse matrix adjustment arising in discriminant analysis," Ann. Math. Statist. 22, 167.

[6]. Brillinger, D. [1964], "The asymptotic behavior of Tukey's general method of setting approximate confidence limits (the jackknife) when applied to maximum likelihood estimates," Rev. Int. Statist. Inst. 32, 202-206.

[7]. Brillinger, D. [1966], "The application of the jackknife to the analysis of consumer surveys," Commentary. The Journal of the Market Res. Soc. 8, 74-80.

[8]. Bromwich, T. J. [1926], An Introduction to the Theory of Infinite Series, 2nd ed., Macmillan and Co. Ldt., New York.

[9]. Burt, J. M., Gaver, D. P., and Perlas, M. [1970], "Simple stochastic networks: Some problems and procedures," Naval Research Logistics Quarterly 17, 439-459.

[10]. Chakrabarty, R. P., and Rao, J. N. K. [1968], "The bias and stability of the jackknife variance estimator in ratio estimation," (abstract) JASA 63, 748.

[11]. Chatterjee, S., and Barcum, S. [1970], "A nonparametric approach to credit screening," JASA 65, 150-154.

[12]. Cochran, W. G. [1968], "Commentary on estimation of error rates in discriminant analysis," Technometrics 10, 204-205.

[13]. Dammkoehler, R. A. [1966], "A computational procedure for parameter estimation applicable to certain nonlinear models of enzyme kinetics," J. Biol. Chem. 241, 1955-1957.

[14]. Das Gupta, S. [1968], "Some aspects of discriminant function coefficients," Sankhyá 30, 387-400.

[15]. Deming, W. E. [1963], "On the correction of mathematical bias by use of replicated designs," Metrika 6, 37-42.

[16]. Dempster, A. [1966], "Estimation in multivariate analysis," Multivariate Analysis, Vol. I, Academic Press, New York.

[17]. Durbin, J. [1959], "A note on the application of Quenouille's method of bias reduction to the estimation of ratios," Biometrika 46, 477-480.

[18]. Fisher, R. A. [1936], "The use of multiple measurements in taxonomic problems," Annals of Eugenics 7, 179-188.

[19]. Fryer, J. [1970], "Jackknifing maximum likelihood estimates in the multiparameter case," (preliminary report)(abstract), Ann. Math. Statist. 41, 1392.

[20]. Gaver, D. P., and Hoel, D. G. [1970], "Comparison of certain small-sample Poisson probability estimates," Technometrics 12, 835-850.

[21]. Gray, H. L., Watkins, T. A., and Adams, J. E. [1972], "On the generalized jackknife, and its extensions, and relation to e_n-transformations, invited paper to appear in Ann. Math. Statist. Feb. 1972.

[22]. Jones, H. L. [1963], "The jackknife method," Proc. of the IBM Scientific Computing Symposium on Statistics, 185-197.

[23]. Kendall, M. G., and Stuart, A. [1961], The Advanced Theory of Statistics, Vol. 2, C. Griffin and Co. Ltd., London, 5-7.

[24]. Kshirsagar, A. M. [1971], Advanced Theory of Multivariate Analysis, Marcel Dekker, Inc., New York.

[25]. Lachenbruch, P., and Mickey, M. [1968], "Estimation of error rates in discriminant analysis," Technometrics 10, 1-11.

[26]. Layard, M. [1971], "Asymptotically robust tests about covariance matrices," to appear in Ann. Math. Statist.

[27]. Layard, M. [1971], "Robust large-sample tests for homogeneity of variances," submitted to JASA.

[28]. Mantel, N. [1967], "Assumption-free estimators using U-statistics and a relationship to the jack-knife method," Biometrics 23, 567-571.

[29]. McCarthy, P. J. [1969], "Pseudo-replication: Half samples," Rev. Int. Statist. Inst. 37, 239-264.

[30]. Miller, R. [1964], "A trustworthy jackknife," Ann. Math. Statist. 35, 1594-1605.

[31]. Miller, R. [1968], "Jackknifing variances," Ann. Math. Statist. 39, 567-582/

[32]. Mosteller, F., and Tukey, J. W. [1968], "Data analysis, including statistics," Handbook of Social Psychology, G. Lindzey and E. Aronson, Eds., Addison-Wesley, Reading Mass.

[33]. Quenouille, M. [1949], "Approximate tests of correlation in time series," J. Roy. Statist. Soc. ser. B. 11, 68-84.

[34]. Quenouille, M. [1956], "Notes on bias in estimation," Biometrika 43, 353-360.

[35]. Raj, D. [1968], Sampling Theory, McGraw-Hill, New York.

[36]. Rao, J. N. K. [1965], "A note on the estimation of ratios by Quenouille's method," Biometrika 52, 647-649.

[37]. Rao, J. N. K. [1969], "Ratio and regression estimators," New Developments in Survey Sampling, N. L. Johndon and Harry Smith, Eds., Wiley-Interscience, New York, 213-234.

[38]. Rao, J. N. K., and Webster, J. [1966], "On two methods of bias reduction in the estimation of ratios," Biometrika 53, 571-577.

[39]. Rao, P. S. R. S. [1969], "Comparison of four ratio-type estimates under a model," JASA 64, 574-580.

[40]. Rao, P. S. R. S., and Rao, J. N. K. [1970], "Some small results for ratio estimators," (abstract), Ann. Math. Statist. 41, 1141-1142.

[41]. Robson, D. S., and Whitlock, J. H. [1964], "Estimation of a truncation point," Biometrika 51, 33-39.

[42]. Salsburg, D. [1971], "Testing dose responses on proportions near zero or one with the jackknife," accepted for publication in Biometrics.

[43]. Schucany, W. R. [1971], "The reduction of bias in parametric estimation," SMU Department of Statistics Technical Report No. 93.

[44]. Schucany, W. R., Gray, H. L., and Owen, D. B. [1971], "On bias reduction in estimation," JASA 66, 524-533.

[45]. Shorak, G. R. [1969], "Testing and estimating ratios of scale parameters," JASA 64, 999-1013.

[46]. Tukey, J. W. [1958], "Bias and confidence in not quite large samples," (abstract), Ann. Math. Statist. 29, 614.

[47]. Walsh, J. E. [1947], "Concerning the effect of intraclass correlation on certain significance tests," Ann. Math. Statist. 18, 88-96.

[48]. Watkins, T. A. [1971], "Jackknifing stochastic processes," Ph.D. dissertation, Texas Tech University.